# Lecture Notes in Networks and Systems

Volume 679

The series "Lecture Notes in Networks and Systems" publishes the latest developments in Networks and Systems—quickly, informally and with high quality. Original research reported in proceedings and post-proceedings represents the core of LNNS.

Volumes published in LNNS embrace all aspects and subfields of, as well as new challenges in, Networks and Systems.

The series contains proceedings and edited volumes in systems and networks, spanning the areas of Cyber-Physical Systems, Autonomous Systems, Sensor Networks, Control Systems, Energy Systems, Automotive Systems, Biological Systems, Vehicular Networking and Connected Vehicles, Aerospace Systems, Automation, Manufacturing, Smart Grids, Nonlinear Systems, Power Systems, Robotics, Social Systems, Economic Systems and other. Of particular value to both the contributors and the readership are the short publication timeframe and the world-wide distribution and exposure which enable both a wide and rapid dissemination of research output.

The series covers the theory, applications, and perspectives on the state of the art and future developments relevant to systems and networks, decision making, control, complex processes and related areas, as embedded in the fields of interdisciplinary and applied sciences, engineering, computer science, physics, economics, social, and life sciences, as well as the paradigms and methodologies behind them.

Indexed by SCOPUS, INSPEC, WTI Frankfurt eG, zbMATH, SCImago.

All books published in the series are submitted for consideration in Web of Science.

For proposals from Asia please contact Aninda Bose (aninda.bose@springer.com).

Phayung Meesad · Sunantha Sodsee ·
Watchareewan Jitsakul · Sakchai Tangwannawit
Editors

# Proceedings of the 19th International Conference on Computing and Information Technology (IC2IT 2023)

Springer

*Editors*
Phayung Meesad
Faculty of IT and Digital Innovation
King Mongkut's University of Technology
Bangkok, Thailand

Sunantha Sodsee
Faculty of Information Technology
King Mongkut's University of Technology
Bangkok, Thailand

Watchareewan Jitsakul
Faculty of IT and Digital Innovation
King Mongkut's University of Technology
Bangkok, Thailand

Sakchai Tangwannawit
Faculty of IT and Digital Innovation
King Mongkut's University of Technology
Bangkok, Thailand

ISSN 2367-3370 ISSN 2367-3389 (electronic)
Lecture Notes in Networks and Systems
ISBN 978-3-031-30473-6 ISBN 978-3-031-30474-3 (eBook)
https://doi.org/10.1007/978-3-031-30474-3



This Springer imprint is published by the registered company Springer Nature Switzerland AG
The registered company address is: Gewerbestrasse 11, 6330 Cham, Switzerland

# Preface

The COVID-19 pandemic has brought about a transformation in our way of life. It has forced us to utilize digital technology to facilitate our daily activities, from dining and education, to communication, work processes, hygiene, and every social activity. It has also significantly affected businesses and industries.

As a sharing platform, the International Conference on Computing and Information Technology (IC2IT) has been organized for 19 years to serve digital communities in exchanging ideas, knowledge, and researches to inspire new digital innovations. Herein, the book presents the main contributions of researchers who participated in the 19th International Conference on Computing and Information Technology (IC2IT2023), held during May 18–19, 2023, in Bangkok, Thailand. The conference proceedings include carefully selected and reviewed research papers in the areas of machine learning, natural language processing, image processing, robotics, as well as digital technology. The submissions received from ten countries were peer-reviewed by at least three members of the program committee. Among these submissions, 18 have been accepted for publication.

With this book, we aim to provide readers with comprehensive their insights into the latest advancements in digital technology. We are confident that they will find it a valuable resource.

Furthermore, the organizing committee would like to extend heartfelt gratitude to the authors, program committee, and collaborating institutions in Thailand and abroad for their support and academic collaboration. Our appreciation also extends to the staff members of the Faculty of Information Technology and Digital Innovation at King Mongkut's University of Technology North Bangkok.

February 2023

Phayung Meesad
Sunantha Sodsee
Watchareewan Jitsakul
Sakchai Tangwannawit

# Organization

## Program Committee

| | |
|---|---|
| S. Auwatanamongkol | NIDA, Thailand |
| T. Bernard | Li-Parad, France |
| S. Boonkrong | SUT, Thailand |
| P. Boonyopakorn | KMUTNB, Thailand |
| V. Chongsuphajaisiddhi | KMUTT, Thailand |
| N. Ditcharoen | UBU, Thailand |
| K. Dittakan | PSU, Thailand |
| T. Eggendorfer | HS Weingarten, Germany |
| M. Hagan | OSU, USA |
| K. Hashimoto | PSU, Thailand |
| K. Hemmi | University of Nagasaki, Japan |
| S. Hengpraprohm | NPRU, Thailand |
| D. V. Hieu | TGU, Vietnam |
| S. Hiranpongsin | UBU, Thailand |
| W. Janratchakool | RMUTT, Thailand |
| C. Jareanpon | MSU, Thailand |
| T. Jensuttiwetchakul | KMUTNB, Thailand |
| W. Jitsakul | KMUTNB, Thailand |
| M. Kaenampornpan | MSU, Thailand |
| M. Ketcham | KMUTNB, Thailand |
| P. Kuacharoen | NIDA, Thailand |
| M. Kubek | FernUni, Germany |
| S. Kukanok | IBA, Thailand |
| P. Kunakornvong | RMUTT, Thailand |
| U. Lechner | UniBW, Germany |
| N. Lertchuwongsa | PSU, Thailand |
| M. Maliyaem | KMUTNB, Thailand |

| | |
|---|---|
| P. Meesad | KMUTNB, Thailand |
| S. Nuchitprasitchai | KMUTNB, Thailand |
| K. Pasupa | KMITL, Thailand |
| S. Plaengsorn | PBRU, Thailand |
| N. Porrawatpreyakorn | KMUTNB, Thailand |
| P. Saengsiri | TISTR, Thailand |
| T. Sarawong | RMUTK, Thailand |
| S. Sodsee | KMUTNB, Thailand |
| W. Sriurai | UBU, Thailand |
| T. Sucontphunt | NIDA, Thailand |
| A. Taguchi | TCU, Japan |
| P. Tangwannawit | PCRU, Thailand |
| S. Tangwannawit | KMUTNB, Thailand |
| C. Thaenchaikun | PSU, Thailand |
| J. Thongkam | MSU, Thailand |
| N. Tongtep | PSU, Thailand |
| K. Treeprapin | UBU, Thailand |
| D. Tutsch | Uni-Wuppertal, Germany |
| N. Utakrit | KMUTNB, Thailand |
| K. Viriyapant | KMUTNB, Thailand |

## Organizing Committee

### *Organizing Partners*

In Cooperation with

King Mongkut's University of Technology North Bangkok (KMUTNB)
Council of IT Deans of Thailand (CITT)
FernUniversitaet in Hagen, Germany (FernUni)
Chemnitz University of Technology, Germany (CUT)
Oklahoma State University, USA (OSU)
Edith Cowan University, Western Australia (ECU)
Hanoi National University of Education, Vietnam (HNUE)
Mahasarakham University (MSU)
Nakhon Pathom Rajabhat University (NPRU)
National Institute of Development Administration (NIDA)
Rajamangala University of Technology Thanyaburi (RMUTT)
Kanchanaburi Rajabhat University (KRU)
Gesellschaft für Informatik (GI)
IEEE CIS Thailand
The Thailand Convention & Exhibition Bureau (TCEB)

# Contents

# A Robust Cursor Activity Control (RCAC) with Iris Gesture and Blink Detection Technique

**Md. Rayhan Al Islam, Maliha Rahman, Md. Rezyuan, Abdullah Al Farabe, and Rubayat Ahmed Khan**

**Abstract** Utilization of computer technology becomes highly difficult and frequently impossible for those with very little or no arm movement. In this study, we suggest an iris-based cursor control system designed specifically for people with physical disabilities. For many years, eye gaze localization has been a particularly prominent area of study. The most sophisticated efforts in this area include Tobi Eyes, Gaze Pointer, and Face Detection using Viola-Jones, although they all have drawbacks including infrared ray health risks and accuracy divergence due to shaky frames. We have created a model to get around these obstacles in our job. In addition, we've designed a hardware design that combines a camera with a wearing cap. Our prototype RCAC makes it easier to execute mouse motions in accordance with the position of the iris. We were 89.8% accurate with our mouse positioning over the user screen during the entire experiment. The entire system was designed for optimal user comfort with the least amount of setup time.

**Keywords** Cursor control · Image processing · Blink detection · Iris gesture · Hough circle transform · Window to viewport transformation · Haar cascade eye trees

## 1 Introduction

One of the essential components for producing computer commands is the cursor. It is conceivable to control the entire computing system using only the human eyes if the iris gesture is accurately measured and used to position the cursor over the proper screen pixel. Worldwide, there are about 5.4 million persons who are paralyzed [1]. Using the flowchart that depicts the system's basic operation, we have demonstrated

Md. R. Al Islam · M. Rahman · Md. Rezyuan · A. Al Farabe (✉) · R. A. Khan
Department of Computer Science and Engineering, BRAC University, 66 Mohakhali, Dhaka 1212, Bangladesh
e-mail: abdullah.farabe@gmail.com

R. A. Khan
e-mail: rubayat.ahmed@bracu.ac.bd



P. Meesad et al. (eds.), *Proceedings of the 19th International Conference on Computing and Information Technology (IC2IT 2023)*, Lecture Notes in Networks and Systems 679,
https://doi.org/10.1007/978-3-031-30474-3_1

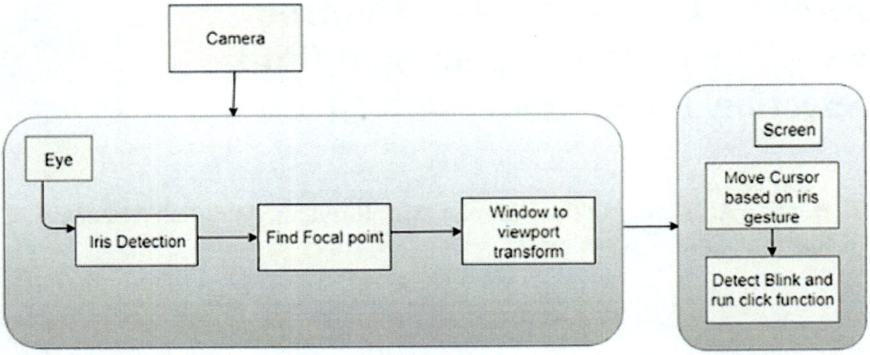


**Fig. 1** Structure of the entire system

how the system truly operates in Fig. 1. The entire system is divided into three parts: video capture; frame pre-processing; focus point recognition; and cursor movement. We identified the iris from the eye and located the focal point, which is utilized to move the cursor on the screen after the camera's footage was captured. To detect blink and execute the click function, the blink detection system is running parallelly. Our motivation is to create a solution that will enable people with physical disabilities to use a computer system just with their eyes, thereby assisting them. As a result, a vision-based approach is considered to be an efficient method for creating human–computer interaction systems. In order to interact with interfaces without the need for a mouse or keyboard, eye movements can be recorded and used as control signals [2]. The project is divided into three sections: cursor control, image processing, and picture capture. It will command the computer mouse to travel to a certain point after the image has been captured using a camera or webcam [3].

## 2 Literature Review

In the existing literature, Yeung, Y. S. et al. [4] have invented a head-mounted framework that will control the mouse cursor of the computer by tracking our eyes. A head tracker has been used to repay head movements during eye tracking. The AsTeRICS has worked as a plugin platform for the software part which will help the whole framework to track eyes through the movement of our heads as well.

In another existing literature, Ramesh, R., and Rishikesh, M. et al. [5] also tried to develop a module to control cursor movement by using a real-time camera and color pointers, Image segmentation, background subtraction, and color tracking are the key points to perform everything that a mouse cursor could do in a computer system.

We have found another existing literature written by S. S. Wankhede, S. Chhabria, and R. Dharaskar. et al. [6] aims at a human–computer interface system by detecting people's eye movements the whole framework could control the mouse moving automatically and through blinking action could simulate mouse-click.

Therefore, after reviewing earlier research, we could see that someone had utilized the ITU gaze tracer, which uses OpenCV and necessitates the configuration of additional expensive hardware as well as head movements for tracking. However, some individuals are unable to turn their heads. and it is true that IR is harmful because it can lead to a number of different skin and eye disorders. Using the segmentation process of viola Jones, which finds and detects eyes, is slow. We have designed and created a model that combines iris recognition image processing methods with hardware design. Our technology is reliable and easy to use, which will assist persons with physical disabilities in overcoming their physical obstacles.

# 3 Algorithms and System Implementation

## *3.1 Frame Processing*

**Determining Region of Interest (ROI):** We have used a cap to place the webcam and detect the region of interest shown in Fig. 2(a). The input video is fragmented into several parts. The Region of interest (ROI) which is the iris portion of the overall frame is cropped from the input frame [7]. In Fig. 2(b) this was the actually captured frame taken from the webcam and Fig. 2(c) was the Region of Interest on which all the algorithm runs.

**BGR to HSV Conversion:** In computer graphics, the RGB color model is presented in a different way called HSV (hue, saturation, value), which reflects the characteristics of color. When comparing this in OpenCV, the ranges must be correctly normalized because those scales vary between different pieces of software [8]. The input frame must first be converted to HSV color in order for our study to successfully detect eyeballs [9]. We have used the built-in BGR to HSV color space conversion method provided by OpenCV.

The reason behind choosing HSV color space is it provides more accuracy in terms of object detection than other available color spaces. In Fig. 3(a), we have shown

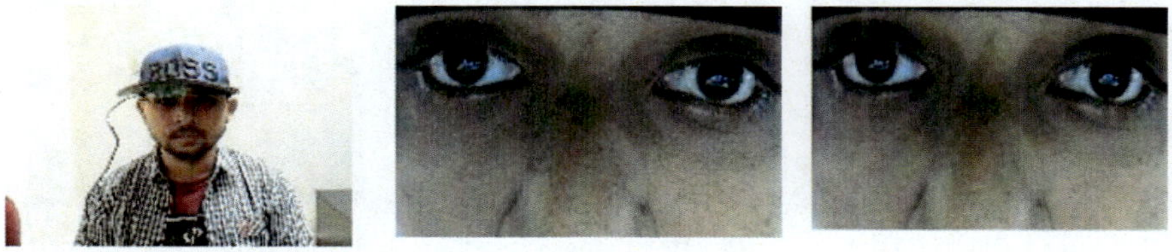

**(a) Hardware Model** **(b) Actual Camera Frame** **(c) Region of Interest**

**Fig. 2** Comparison between camera frame and Region of Interest (ROI) the before version of the conversion from BGR to HSV and, in Fig. 3(b), we can see that the color space is converted to HSV in order to make the iris portion distinct and the other portion less visible. The masking process clears out the unnecessary frames and makes the region of interest more visible because of planning to use only the iris portion and run our algorithm over the rounded portion of our eyes only [10]. Therefore, using the masking we filtered the black iris part and made it white with the bitwise method. Other than the iris part, the complete frame was darkened so that our algorithm does not process any kind of noise in the frame region [11]. Figure 3(b) shows the frame's condition prior to masking, and Fig. 3(c) shows the frame's condition following masking. After that, the sharpness will be improved and the noise will be eliminated, and we will then conduct additional functions.

**Averaging Blur:** From image processing average blurring is a type which is image smoothing method that helps to reduce noise by blurring them out. When using an average blurring filter, each pixel in the image is taken into account individually. The kernels, which move along with the pixels being filtered, are the groups of pixels [12]. An image's "outlier" pixels, which could be noise, are removed by blurring it. An example of using a "low-pass" filter on an image is blurring. In Fig. 4(a) we showed the frame before using the blur filter and in Fig. 4(b) the frame using the averaging blur filter.

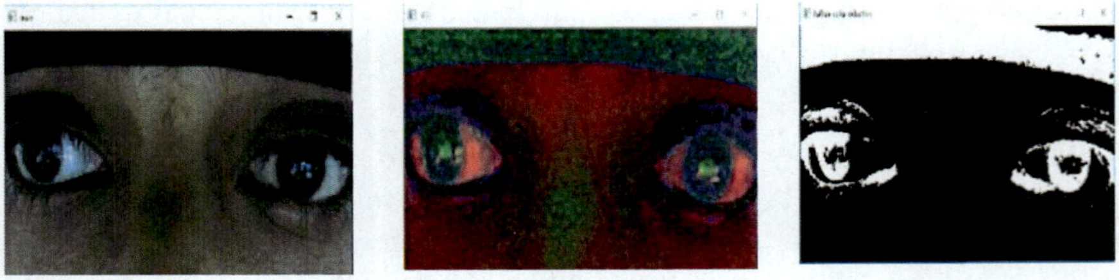

**(a) Before BGR to HSV Conversion** **(b) After BGR to HSV Conversion** **(c) After Performing Masking**

**Fig. 3** Before and after result of BGR to HSV conversion

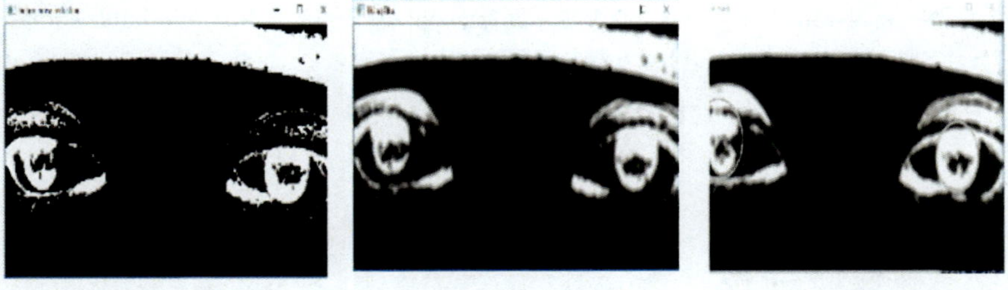

**(a) After Using Gaussian Blur** **(b) After Using Averaging Blur** **(c) Applying Hough Circle Transformation**

**Fig. 4** Comparison between Gaussian blur and averaging blur

## 3.2 Algorithms

In the detection (One-Eyed Fixation) process, we applied the Hough circle approach to the OpenCV blur function's smooth image. A NumPy Nd-array holding the circle's center coordinates is returned by the Hough circle. The left and right eyes were then identified using the array value. Our program will change the position of the other eye and calculate the pupil distance if one eye cannot be located. The system's real viewport is then determined from the position of the iris circle. CV is the hough circle method. Image, circles, method, dp, minDist, param1 = 100, param2 = 100, minRadius = 0, and maxRadius = 0; this is a built-in property of OpenCV.

**Hough Circle Transform:** Hough circle transform helps to find circular items in a digital image. Equation 1 describes a circle on two-dimensional surfaces.

$$(x - a)^2 + (y - b)^2 = r^2 \tag{1}$$

The circle's center is shown as (a, b) and its radius is shown as r. The image frames are first processed using Gaussian blur, and then clever edge detection is carried out. The process of detecting and voting the circles is then checked. The variables (a, b, and r) are all initially set to 0. Equations 2 and 3 are then used to calculate the polar coordinate of the center.

$$a = x - r \times cos\left(\frac{t \times \pi}{180}\right) \tag{2}$$

$$b = y - r \times sin\left(\frac{t \times \pi}{180}\right) \tag{3}$$

The returned result where [a, b] denotes the location of the circle's center and [r] denotes the radius. The image in Fig. 4(c) depicts the two Hough circles that were identified. The canny edge is represented by parameter one. Low value detects more edge while high value detects less. We have entered the number 370 to solely calculate the rounded iris edge. The param2 parameter specifies the circular center's accumulator threshold value. More bogus circles will be picked up as it becomes smaller. Twenty have been added to our case. We have established a minimum radius of 40 and a maximum radius of 50 for the localization of the iris [13]. We can skip erroneous or incorrect circles in the frame thanks to this argument. According to the specified criteria, the mentioned image in Fig. 4(c) displays the two Hough circles that were discovered. The coordinates of the two iris circles that were found are provided in the numpy nd-array that was returned. For ***Data Processing,*** we determine the value of the circles in a NumPy array form after applying the Hough Circle Algorithm. We are creating an array that includes both circle's x and y coordinates of the center of the circles. Then for **Iris Adjustment,** we obtain the left and right eye values in our system after detecting the circles using the Hough Circle technique.

**One-Eyed Fixation:** Following the discovery of the circles one eye loses focus. As a result, the circle detection algorithm returns just one circle to us. Furthermore, since we are averaging the midpoints of both circles, the center value of only one circle cannot accurately represent the mouse position. Therefore, we have suggested a different method that we have dubbed "OneEyedFixation".The equations are provided.-

$$\text{DistanceOfBothEye} = \text{RightX - LeftX}$$
$$\text{RightX} = \text{LeftX} + \text{DistanceOfBothEye}$$
$$\text{LeftX} = \text{RightX} - \text{DistanceOfBothEye}$$

To do this, we must first determine which eye is being detected. The ROI frame's position is fixed at (0, 0), therefore the left eye can only display inside a specified range. We can examine the left-eye appearance pixel range because our frame is constant across all users. The detected circle is the left eye if the numbers are within the range; otherwise, it is the right eye. We can make our pointer go smoothly and avoid all types of mistake positions by employing the two suggested approaches. For **Window to Viewport Transformation,** the pointer is positioned on the monitor screen via a window-to-viewport transformation. The center of the iris determines exactly where our cursor will be. Finding the area that our iris covers were the first step. The midway of our iris can only travel a certain amount of space. The range or size of our window would be between our eyes because we are averaging the two irises.

**Calibration:** The process of calibrating involves defining some unknown values in relation to known values. Our program's four windows will first be visible for five seconds, displaying four separate points at the corners of the screen. The user looks at each corner's point while it is on display for a predetermined amount of time. Here, the x-axis and y-axis minimum and maximum values are determined by the points in the upper left and lower right corners, respectively. In the meantime, the window's coordinates are saved as the average values of both eye irises' midpoints when the user stares at the four spots. The upper-right pointer window and the lower-left pointer window appear after the upper-left pointer window disappears. We quickly identified the location of our particular window using this calibration. The size of the real window that the eye's focal point goes through is depicted in Fig. 5.

**View Port Determination:** Our system's viewport fluctuates because different display sizes are available to users. Our technology initially calculates the window size using the calibration method before measuring the device display's width and height. For **Cursor Movement Control,** on a computer screen, lines can be drawn between any two pixels using the midpoint line drawing algorithm. We carefully selected this approach since it is relatively uncomplicated and only calls for integer values and straightforward computations, which is exactly what we need to move the mouse around the entire computer screen randomly. This algorithm is more time-efficient because it doesn't use multiplication or division. Instead of constructing

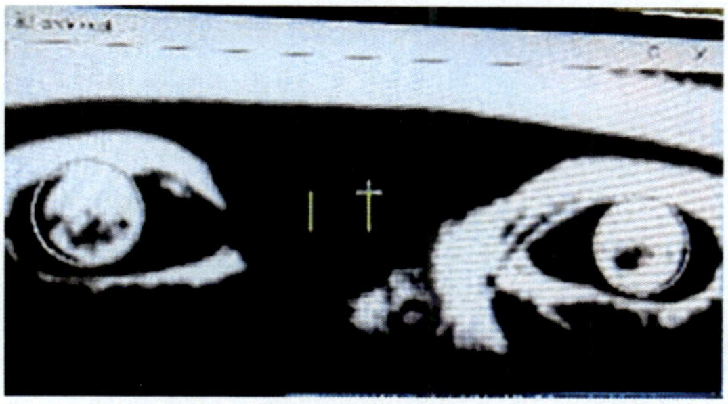

**Fig. 5** Window of the cursor

lines for our system, we simply shifted the pointer from one spot to another. Using Eqs. 2 and 3, we determined the FocusPoint (x, y) of our eye after obtaining the relative center positions of the left and right eye in the viewport. The primary focal point is the intersection of the centers' x- and y-coordinates. The cursor starts off in the (0, 0) location. To shift the cursor smoothly from one location to another, we kept both the previous and next pixel values. As the focus point values can be in a floating condition after calculation, before the cursor movement algorithm enters, we first had to round up the cursor positions. These zones cover every side of the computer screen. We can move the cursor in each direction accurately. Next step for **Blink detection for Click Function,** closed eyelids cannot be detected by the integrated Haar-cascade. As a result, we have run the click function based on the classifier's detection of the closed eye. The left-click function will operate for the closed left eye, and the right-click function will operate for the closed right eye. To identify eye regions, we have used the Haar-cascade eye tree eyeglasses classifier. A classifier called Haar-cascade detects a variety of items because it contains several trained sets. The eyes classifier is what we are employing. The inability of this classifier to recognize closed eyelids is one of its properties. As a blink detector, we are recognizing the closed-eyes condition.

## 4 Result Analysis

**Evaluate the Accuracy of Cursor Movement and Cursor Pointer:** We chose three different activities to do in order to assess the cursor's accuracy. In scenario 1, the user must first choose a folder from the desktop before deleting a file inside. In scenario 2, the action that will be taken is to launch Windows Defender and conduct a short scan. In scenario 3, the user must continuously refresh twice. Three demographic groups—young adults (20–40), seniors (65+), and teenagers (13–19) were used in our tests (40 and above). After that, we recorded how long it took them to finish each activity. In each group, we took 10 people (a total of 30) for the testing. After that, we stored the number of seconds they took while completing the tasks. The same tasks performed by all groups took a diverse amount of time. In order to perform case 1, teenager group took around 6.87 s, adult group took around 6.5 s and the old group took 7.5 s. We can see that, adults took less time than the other two groups. Using the mouse, we recorded the values of a few selected spots. These points were chosen utilizing several desktop-based folders. For this examination, 10 values were collected. For instance, if a folder is located on the desktop in the left-hand corner of the monitor, using the mouse we select the folder at pixel (18, 1074), and using our system we can select the folder at pixel (17, 1065). Equation 4 can be used to determine the accuracy from the obtained coordinate values the accuracy for the X-axis becomes 91.7%, and for the Y-axis accuracy becomes 90.8%.The complete HD screen was divided into pixels (1920*1080). As a result, the folder position was calculated using screen pixels. Let's say the folder is located at (18,1074) on the x and y axes, and utilizing eye movement, the pointer reached (17,1065), which is nearly exact. TrueValue measures how closely a measured value resembles a true or standard value. The true value of the property being measured is different from SystemValue, which is the measurement value provided by the measuring device.

$$Accuracy = \{(TrueValue - SystemValue)/TrueValue\} \times 100 \tag{4}$$

## 4.1 Accuracy of Blinks in Terms of Distance

We used the click function while maintaining various distances to assess the blink accuracy in terms of distance. In order to achieve our goal of 15 perfect clicks, we had to click a little bit more at a 2-foot distance around 17 times. Table 1 displays the results with accuracy at various distances. Since the camera is fixed in place, distance is irrelevant in our system.

**Table 1** Accuracy of blinks in terms of distance

| Distance | No. of clicks | No. of attempt | Accuracy |
|---|---|---|---|
| 2.5 ft | 15 | 18 | 83.2% |
| 2.0 ft | 15 | 17 | 88.23% |
| 1.5 ft | 15 | 17 | 88.23% |
| 1 ft | 15 | 16 | 93.75% |

**Table 2** Comparison between existing system and our system

| Functionalities | Gaze pointer | Our system |
|---|---|---|
| Distance flexibility | N/A | Available |
| OS independent | N/A | Available |
| Screen size independent | Available | Available |
| Detection over eye-glass | Available | Less Accurate |
| Calibration process | Lengthy | Short and simple |
| One eye usability | N/A | Available |

### *4.2 Comparison with Existing System*

Webcam eye-tracking software called GazePointer enables you to control the mouse cursor with your eyes. We conducted a comparison with the previously existing system in order to demonstrate the improvements in our system in Table 2. Python is already installed on the machine, this system will function there. Both systems are agnostic of screen size, therefore whether a 1 K or 2 K display is used, the system will function as intended.

## 5 Conclusion

The most recent developments in technology provide us with an opportunity to introduce a system that is cost-effective. This cursor control technology will open up new accessibility possibilities for disabled people. However, the accuracy rate for recognizing closed eyelids fluctuated when the light was not very bright. Therefore brightness might be a constraint. By employing a more sophisticated camera with a precise focus, the problem can be solved. Additionally, Gaze Pointer can function while wearing spectacles, although our device performs less well in this area. Our system's restriction is caused by the low-quality camera, which is unable to eliminate the glass reflections above the eyes. The user needs to spend a few seconds calibrating the device because all he needs to do is utilize the calibration photographs to determine the window of his iris, which we mentioned in Fig. 5. In the future by demonstrating its accuracy in various settings with users of all ages, the system can enhance human–computer interaction. We will need to use a better webcam in the future if we want to increase our accuracy. Therefore, with increased camera focus, our system can get around this restriction and produce superior results.

## References


1. C.D.R. Foundation Paralysis statistics. Stats about paralysis. https://www.christopherreeve.org/living-with-paralysis/stats-about-paralysis#:~:text=According%20to%20the%20study%2C%20there,higher%20than%20previous%20estimates%20showed. Accessed 21 Nov 2022
2. Poole A, Ball LJ (2006) Eye tracking in HCI and usability research. In: Encyclopedia of Human computer interaction, pp 211–219
3. Salunkhe P, Patil AR (2016) A device controlled using eye movement. In: 2016 international conference on electrical, electronics, and optimization techniques (ICEEOT), pp 732–735. IEEE
4. Yeung YS (2012) Mouse cursor control with head and eye movements: a low-cost approach. Master's Thesis, University of Applied Sciences Technikum Wien
5. Ramesh R, Rishikesh M (2015) Eye ball movement to control computer screen. J Biosens Bioelectr 6(3):1
6. Wankhede S, Chhabria S, Dharaskar RV (2013) Controlling mouse cursor using eye movement. Int J Appl Innov Eng Manag 36:1–7
7. CVisionDemy Extract ROI from image with python and opencv. Extracting a ROI (Region of Interest) using OpenCV and Python. https://cvisiondemy.com/extract-roi-from-image-with-python-and-opencv. Accessed 21 Nov 2022
8. Introduction to OpenCV-Python Tutorials — OpenCV-Python Tutorials beta documentation. (n.d.a). https://opencv24-python-tutorials.readthedocs.io/en/latest/py_tutorials/py_setup/py_intro/py_intro.html. Accessed 21 Nov 2022
9. Manjare S, Chougule SR (2013) Skin detection for face recognition based on HSV color space. Int J Eng Sci Res Technol 2(7):1883–1887
10. X. S. Ltd Image masking. http://www.xinapse.com/Manual/masking.html. Accessed 21 Nov 2022
11. Priyanka S, Kumar N (2016) AS noise removal in the remote sensing image using kalman filter algorithm. Int J Adv Res Comput Commun Eng 5:894–897
12. S.C. Foundation Blurring images (2016–2018). https://mmeysenburg.github.io/image-processing/06-blurring/. Accessed 21 Nov 2022
13. O.D. Team Hough circle transform. Hough Circle Transform. https://docs.opencv.org/2.4.13.7/doc/tutorials. Accessed 21 Nov 2022

# Lesion Detection Based BT Type Classification Model Using SVT-KLD-FCM and VCR-50

**Fathe Jeribi and Uma Perumal**

**Abstract** Due to delayed diagnosis and treatment, Brain tumors have a very low survival rate and are easily detected on MRIs. With the advancement of science and technology, ML models attempted to identify and classify BT diseases. However, the models have constraints on lesion identification and accurate region prediction. Hence, to overcome these constraints and other problems in the way to predict BR disease, a novel deep learning model has been proposed. The input MRI dataset has been taken for this paper, and pre-processed for enhanced accuracy at the segmentation stage. Pre-processing includes contrast enhancement, edge enhancement using ACMPO, and Skull removal. Pre-processed results are segmented using SVT-KLD-FCM. Segmentation output was two clusters with and without lesions. Without lesion images were declared as normal, and with lesions are given for feature extraction and then to classifier VCR-50. Classification results in three types of classes namely glioma, meningioma, and pituitary. The proposed model was implemented and achieved accuracy of about 98.77% better results than previous models.

**Keywords** Lesion identification · Alpha Channel Masking-Prewitt Operator (ACMPO) · SC with Voxel Threshold-Kullback Leibler divergence –Fuzzy C Means Segmentation algorithm (SVT-KLD-FCM) · Vector map Convolution-ResNet-50 (VCR-50)

## 1 Introduction

BT arises due to the abnormal growth of cells in an uncontrollable manner. BTs are divided into primary and secondary tumors [1]. People can suffer without understanding the risks of BT since its cause is unknown and its symptoms are unclear

F. Jeribi (✉) · U. Perumal
College of Computer Science and Information Technology, Jazan University, Jazan, Saudi Arabia
e-mail: fjeribi@jazanu.edu.sa

U. Perumal
e-mail: uperumal@jazanu.edu.sa

 
P. Meesad et al. (eds.), *Proceedings of the 19th International Conference on Computing and Information Technology (IC2IT 2023)*, Lecture Notes in Networks and Systems 679,
https://doi.org/10.1007/978-3-031-30474-3_2

[2]. Moreover, based on the severity, the tumors are dived into benign and Malignant [3]. Benign is not cancerous, while malignant is cancerous. Some of the malignant tumor types are glioma, meningioma, and pituitary. Hence, detecting them is important to start proper treatment [4]. Many medical imaging modalities are being utilized to detect brain anomalies i.e. computed tomography, Magnetic Resonance Imaging (MRI), etc. [5]. However, the MRI scan is considered as better because it is particularly efficient when it comes to imaging soft tissues and BT growth imaging [6].

In the early days, clinical experts analyze MRI images for tumor, which is a challenging process; also a tumor cannot be detected in every region with the naked eye. Therefore without human intervention, diagnosis of tumors require the use of automated systems [7]. Various deep learning algorithms, such as Deep Neural Networks, Convolutional Neural Networks (CNNs) are introduced for BT detection (BTD) and classification. To improve the accuracy of the models, the images are first preprocessed followed by tumor segmentation. For segmentation, various techniques were utilized, such as grabcut [8] and graph signal processing [9]. But, these existing works of BT tumor detection contain limitations, such as increased time complexity; also, neither the tumor boundary nor its intensity are clearly attained [10]. Thus, to overcome these limitations, VCR50-based BT classification is proposed in this paper.

## *1.1 Problem Statement*

The existing models for BTD contain certain limitations, which are stated as follows:

- Segmentation algorithms in existing works resulted in the drawback of over-splitting images.
- During BT classification, previous algorithms do not take spatial consistency (SC) information for classification.
- While using Segmentation for both region identification and type detection, the system becomes highly complex.

To overcome these issues, a model is proposed with the following contributions:

- After the identification of lesions, the segmentation performed using SVT-KLD-FCM.
- For improved segmentation accuracy, edge enhancement was performed using ACMPO.
- For low complexity in BTD, VCR-50 based classification has been proposed.

The rest of this paper has been organized as: Sect. 2 discusses the related works and their limits, Sect. 3 describes the proposed control model, Sect. 4 presents the experimental results, and finally, Sect. 5 concludes the paper.

## 2 Related Works

K-Nearest Neighbor classifier model for the classification of BTs, The tumor regions are segmented with the Optimal Possibility FCM clustering algorithm [11]. The experimental results showed better tumor detection of the presented method. But, more time would be taken to produce clusters due to the performed centroid selection.

In [12], the authors investigated a hybrid threshold approach. This model combined the CNN and Support Vector Machine (SVM) for the classification. The experimental analysis revealed a higher accuracy level for the CNN-SVM classifier. Yet, the effect of the skull background was not considered that reduce the accuracy of the classifier.

A model developed for the detection of the multi-class tumor using MRI by CNN model [13]. It achieved higher classification accuracy without handmade features. Still, the CNN could not encode the position of objects, which affected the effective tumor detection.

A BTD with MRI deep hashing was introduced by the authors [14]. The hashing combined interpretability and feature fusion to recover the low image resolution. Based on the experiment results, the model could identify tumor regions. However, the distribution of pixels in the MRI led to incorrect detection.

The authors demonstrated a Long Short-Term Memory (LSTM)-based model for BTD [15]. In LSTM, hidden units are optimally selected after performing extensive experiments. The results confirmed that radiologists could accurately classify BTs using the method. But, the thresholding led to misclassification.

An improved framework for BT analysis with MRI images was developed [16]. It uses You-Only-Look-Once-version2 (YOLOv2)-inceptionv3 model. The proposed method achieved better prediction scores in tumors localization. However, the model caused an overfitting problem without appropriate pre-processing.

Deep-CNN (DCNN) for BT classification was suggested by the authors [17]. In the DCNN, a non-linearity layer was added to ameliorate the fitting ability of CNN. Results of the experiment indicated that the DCNN model was effective. Yet, the overall classification accuracy of the DCNN was lesser than the existing works.

## 3 Research Methodology

BTD is challenging because the growth of small cells gets abnormal growth and prediction is not possible. Hence, a novel model is proposed based on the deep learning approach and shown in Fig. 1.

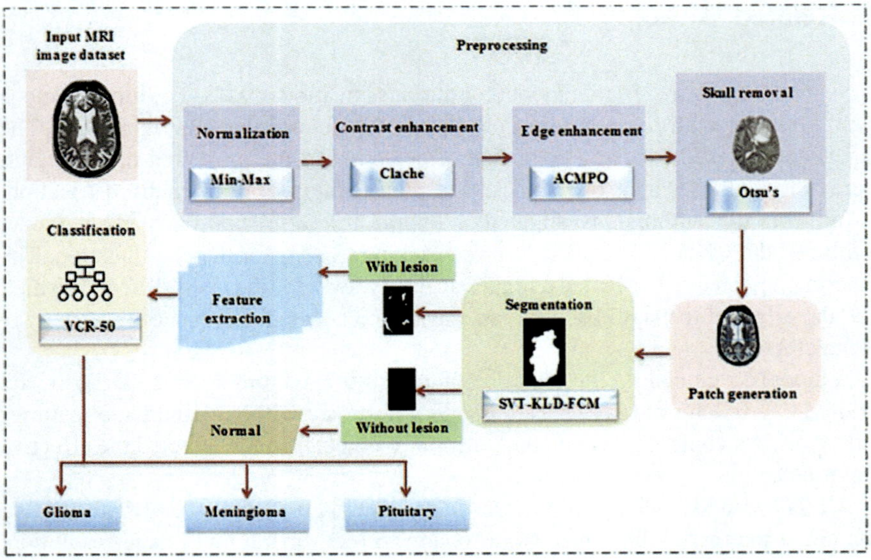


**Fig. 1** Block diagram of proposed model

## 3.1 Pre-processing

The input MRI image ($I_{MRI}$) has been taken from a publically available dataset and the process started with pre-processing the image, which is discussed below:

### 3.1.1 Normalization

Image ($I_{MRI}$) is sub-sampled to reduce the background and then, given to Min–Max normalization. Image data has been normalized in the range of [0, 1]. The Transformation function has been expressed as:

$$\overline{e} = \frac{e - e_{\min}}{e_{\max} - e_{\min}} \tag{1}$$

Here, $e$ represents the initial data, $e_{\min}e_{\max}$ represents the maximum and minimum data. $\overline{e}$ denotes the normalized image.

### 3.1.2 Contrast Enhancement

The normalized image ($\overline{e}$) was then given for contrast enhancement using Contrast Limited Adaptive Histogram Equalization (CLAHE) for its prevention of amplification of noise. Here, the pixels are redistributed and given as:

$$\xi = \left(\frac{G_{avgclip}}{G_{remain}}\right) \tag{2}$$

Here, $\xi$ denotes the positive integer value that was greater than or equal to 1. The remaining number of clipped pixels was denoted as $G_{remain}$ and clipped pixels were summarized as $\left(G_{avgclip}\right)$. The density of each intensity value $(D_i)$ based on Rayleigh forward is termed as:

$$D_i = D_{\min} + \sqrt{2\lambda^2[-\ln(-\psi(i))]} \tag{3}$$

where, the lower bound of the pixel value is $D_{\min}$ and $\lambda$ is Rayleigh scaling parameter, and $\psi(i)$ is the transfer function. A higher $\lambda$ value will improve the contrast enhancement in the image. The enhanced image thus obtained is denoted as $e_{CE}$.

### 3.1.3 Edge Enhancement

The contrast enhanced image might have disturbed edges that might reduce the segmentation accuracy. Hence, edge enhancement is performed using ACM-PO. The conventional Prewitt operator (PO) has been selected to overcome the problem of disturbed edges and also detect the edge orientations, still, to improve the edge weights, an Alpha channel masking (ACM) has been included. The working of this operator is convolved with three masks, horizontal mask $(H_m)$, vertical mask $(V_m)$ and ACM mask $(M_{acm})$.

$$H_m = \begin{bmatrix} 1 & 0 & -1 \\ 1 & 0 & -1 \\ 1 & 0 & -1 \end{bmatrix} \tag{4}$$

$$V_m = \begin{bmatrix} 1 & 1 & 1 \\ 0 & 0 & 0 \\ -1 & -1 & -1 \end{bmatrix} \tag{5}$$

$$M_{acm} = \begin{bmatrix} -1 & -1 & -1 \\ 0 & 0 & 0 \\ -1 & -1 & -1 \end{bmatrix} \tag{6}$$

After performing this masking, the results were obtained with enhanced edges and undisturbed edges $(e_{ee})$.

### 3.1.4 Skull Removal

In previous works, the skull was removed from the image and given for image enhancement, but still, cerebrospinal fluid (CSF) space was partially preserved, as it limits accuracy. Hence, hereafter, edge enhancement of the skull removal process takes place. For this purpose, Otsu's Historical Equalization (HE) method has been adopted. The Image $(e_{ee})$ is given for the HE equation. Erosion was to eliminate meninges and CSFin $(e_{Th})$ by employing structural elements namely $A\, and\, B$.

$$A \Theta B = \{T | (B)_T \subseteq A\} \quad (7)$$

Here, $\Theta$ is the morphological operation, and $T$ is displacement function with all sets of points in $(e_{Th})$. On the other hand, dilation is performed to remove intracranial tissues and given as:

$$A \oplus B = \{T \in B_T \cap A \neq \phi\} \quad (8)$$

Here, $\oplus$ is the dilation function and $\phi$ is the overlapping element set point. The final skull removed image is denoted as $e_{SR}$.

## 3.2 Patch Generation

Before segmenting the image $e_{SR}$, patch generation takes place. Patch-based image segmentation was done using the non-local mean value obtained from the image. In this patch generation stage SC of the image (i.e. appearance of parts of the brain with different brain images at the same region) has been used as a metric for localization of patch generation. The resultant patch generation was denoted as $e_{SR_i}$.

## 3.3 Segmentation

Segmentation image $e_{SR_i}$ was done using SVT-KLD-FCM. The conventional Fuzzy C Means algorithm has been selected for its flexibility over image points Still, to improve the proposed model, the decision on Fuzzy is modified with SC with Voxel Threshold-KLD algorithms. Figure 2 illustrates the flow chart of this algorithm.

To find the region of interest (ROI) and set up a bounding box, SC with Voxel Threshold has been used. The Lesion mask was generated and the final set of candidate regions consisting of all Voxel values above the spatial intensity in the consistency region are given as follows:

$$Lesion\, mask(e_{SR_i}) = \begin{cases} 1\ Sp(e_{SR_i}) > VT \cap e_{SR_i} \\ 0\ Otherwise \end{cases} \quad (9)$$

Here,$e_{SR_i}$ denotes the obtained patch wise images $i = 1, ...n$, and $n$ is the last patch of the image. $Sp(.)$ is the spatial intensity value of the image patches. After masking, the

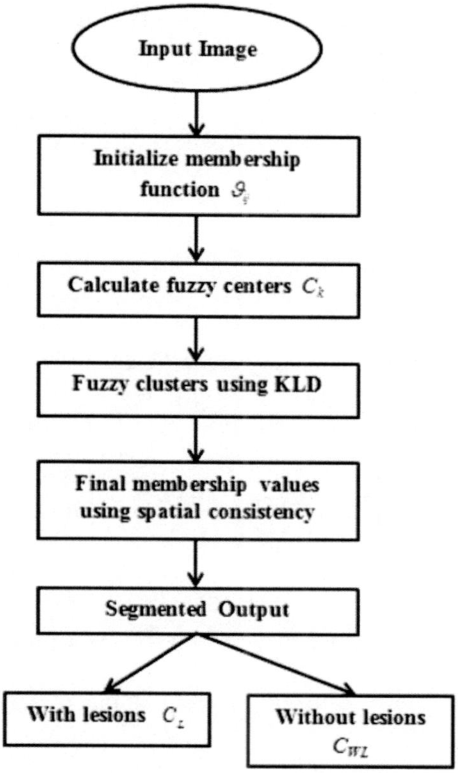


**Fig. 2** Flow chart of SVT-KLD-FCM

intensity information with a weighting parameter $\delta$ is used to determine the relative importance of intensity between patches and select the perfect lesions.

$$\delta = \delta_0 * (t - 1) \tag{10}$$

Here $\delta_0$ is the denotation for label metric and intensity metric at the same scale and the iteration number denoted as $t$. $*$ is the multiplication function. The steps of this segmentation process are given below

***Step 1:*** Initialize the cluster center $c_k$ and set the iteration count $t_i = o$, where $k$ the number of clusters.

***Step 2:*** Initialize fuzzy partitions

$$\vartheta_{ij} = \frac{1}{\sum_{f=1}^{F} \left( \frac{\|e_{SR}(i) - c_k\|}{\|e_{SR}(j) - c_k\|^{2/(k-1)}} \right)} \tag{11}$$

where,$\vartheta_{ij}$ is the fuzzy partition result, $e_{SR}(i)$ is the integral values of image $e_{SR}$, and the orientation values of the image $e_{SR}(j)$. $f$ is the density value that ranges from $1\ to\ F$.

***Step 3:*** Increment $t_i = t_i + 1$ and calculate $c_k$ for all clusters using

$$c_k = \frac{\sum_{j=1}^{N} \vartheta_{ij} e_{SR}(j)}{\sum_{j=1}^{N} \vartheta_{ij}} \tag{12}$$

$j$ is the eigenvector values that range from $1\ to\ N$.

***Step 4:*** Update $\vartheta_{ij}$ using (12)

***Step 5:*** For each pixel, the density of kernels was estimated using KLD as:

$$\overline{g(\vartheta_{ij}, c_k)} = \frac{1}{N} \sum_{i=1}^{N} \frac{1}{h^{\delta}(H_i^{\delta})^j} c_k^{\delta} \times \left(d\left(c_{k^{\alpha}}, e_{SR}(i), H_i^{\delta}\right)\right) \times c_k^{\delta}\left(\left\| c_k - e_{SR}(i)/h^{\delta} H_i^{\delta} \right\|^2\right) \tag{13}$$

where distance $d\left(c_{k^{\delta}}, e_{SR}(i), H_i^{\delta}\right)$ is defined as:

$$d\left(c_{k^{\delta}}, e_{SR}(i), H_i^{\delta}\right) = (e_{SR}(i) - c_k)^T H_i (e_{SR}(i) - c_k) \tag{14}$$

The number of data vectors in clusters is denoted as $H_i^{\delta}$ and $h^{\delta}$ is the unstable clusters after cluster center selection.

***Step 6:*** Apply variances at diagonal items and the mean shift was iterated and consider the previous position. Merging pixels with less distance and repeat the steps until maximum iteration.

The results will be labeled as clusters that are with lesions $(C_{WL})$ and without lesions $(C_L)$. Here, lesions that are not considered as normal $(Nor)$ and with lesions are given to the classifier.

### 3.4 Feature Extraction

From the clustered result,$(C_{WL})$ was considered for feature extraction. Features extracted are Discrete Wavelet Transform (DWT)$\left(f_1^m\right)$, Histogram of Oriented Gradients (HOG)$\left(f_2^m\right)$, Gray-Level Co-occurrence Matrix (GLCM)$\left(f_3^m\right)$, Local Tetra Pattern (LTP)$\left(f_4^m\right)$, and shape features $\left(f_5^m\right)$. It was overall features $f^m$ are expressed as:

$$f^m = \left\{f_1^m, \ldots .f_5^m\right\} \tag{15}$$

### 3.5 Classification

Classification of BTS from the features extracted was done using the VCR-50 classifier model. The conventional ResNet-50 model was selected for its superior performance on brain tumor images and still, it has the disadvantage of slow convolution. Hence, to overcome this issue transposed convolution is modified with Vector Map Convolution (VMC). VCR-50 consisted of Convolution Layers (CL), max-pooling layers, dropout layers, and fully connected layers activated with the softmax function.

Initially, the extracted selected features are given as input to the CL. The feature detectors are the array of weights that represents the part of the input data. Here VMC $(V_{mc})$ is expressed as:

$$V_{mc}(F_n, \psi) = \oint \left|\frac{1}{|\Phi_i(\psi)|}\right|^2 \sum_{F_n(i,j)\in\Phi_i(\psi)} F_n \tag{16}$$

where, $\psi$ is the kernels from the output of CL $(t \times t)$ matrix, $F_n$ denotes the activation function, $\Phi_i(\psi)$ denotes the kernel's sub domain, and $F_n(i, j)$ denotes the coordinate of the point $n$. During testing, matching is performed to detect discriminating *features in* an image and find the disease of the corresponding input image namely glioma, meningioma, and pituitary.

**Input:** Extracted features $\left(f^{m}\right)$

**Output:** Classified output $\left(O_{c}\right)$

**Begin**

**Initialize** VCR-50 layers, the maximum number of iteration $\left(k^{\max}\right)$

**Set** $k = 0$

**While** $k \le k_{\max}$

**Input** $f^{m}$ to the convolution layer

**If** $1 \le m \le M$

**Perform** Vector map convolution $\left(V_{mc}\right)$

$$V_{mc}(F_n, \psi) = \oint \left| \frac{1}{| \Phi_i(\psi) |} \right|^2 \sum_{F_n(i,j) \in \Phi_i(\psi)} F_n$$

**Obtain** Convolution output

**Evaluate** pooling output

**Determine** fully connected layer output

**End if**

**Set** $k = k + 1$

**End while**

**Return** Classified output

**End**

**Algorithm 1:** Pseudo code for VCR-50

## 4 Result and Discussion

Using state-of-the-art methods and existing approaches, the proposed approach is experimentally verified for reliability. The experiments are performed on the working platform of MATLAB. The sample image results on the BT MRI dataset are given in Fig. 3.

### *4.1 Dataset Description*

The performance of the methodologies is analyzed on the publically available BT MRI dataset. This database contains 7022 images of brain MRIs, organized into four categories: glioma-meningioma-no tumor, and pituitary. 80% and 20% of the data are utilized for the training and testing process.

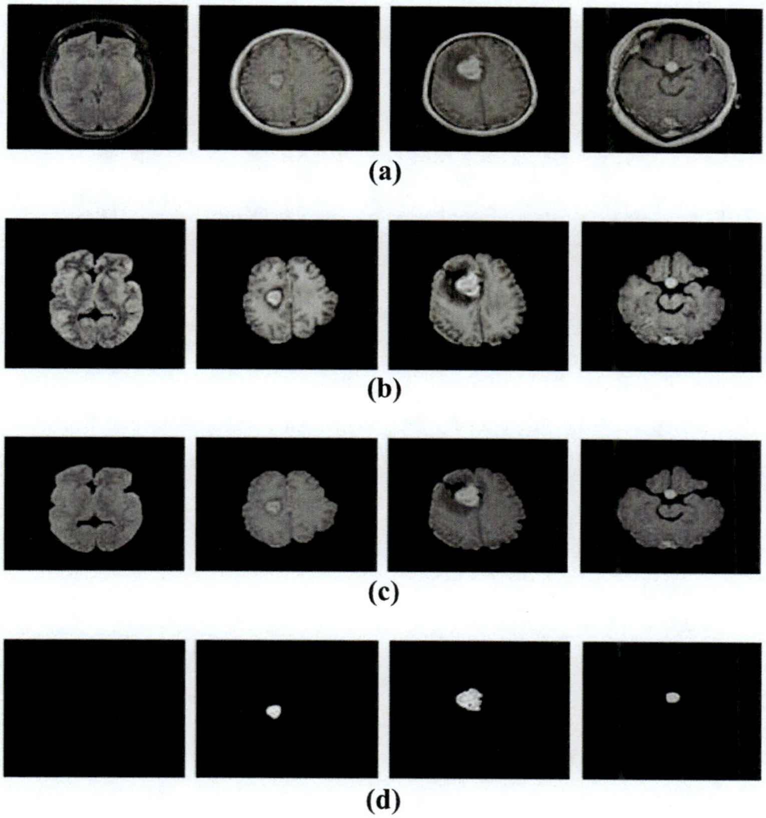

**Fig. 3** Sample Image Results of (a) Input Image (b) Edge Enhanced Image (c) Skull Removed Image and (d) Segmented Image

## *4.2 Performance Analysis of Classification*

Here, the accuracy, recall, precision, F-measure, and training time experimental results of proposed VCR-50 and baseline ResNet-50, AlexNet, DenseNet, and CNN are analyzed.

In Fig. 4, it can be seen that with the VC technique in ResNet-50, the accuracy, recall, precision, and f-measure levels are enhanced by 2.82%, 3.15%, 3.65%, and 3.40% than the ResNet-50 classifier. Also, the proposed classifier attained better results than the AlexNet, DenseNet, and CNN classifiers. This shows that the classified results of the proposed VCN-50 are more reliable for BTD.

In Fig. 5, it is confirmed that the ResNet-50 takes 6498 ms, 27010 ms less time than the AlexNet and CNN. But, the proposed VCR-50 takes 7212 ms less training time than the ResNet-50, which shows the time efficiency of the proposed VCR-50 BT classifier.

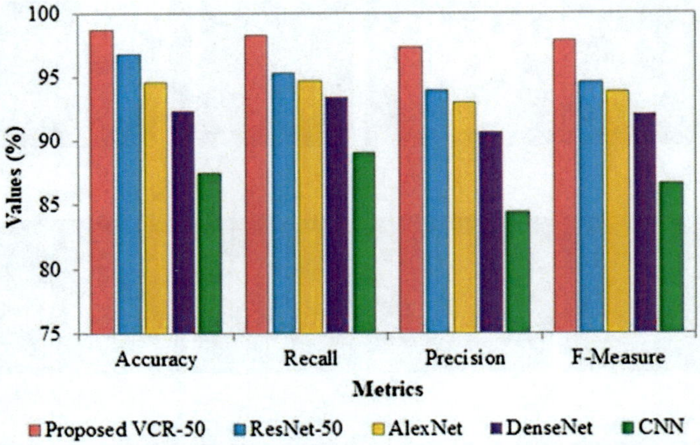


**Fig. 4** Accuracy, precision, recall, and F-measure performance of classifiers

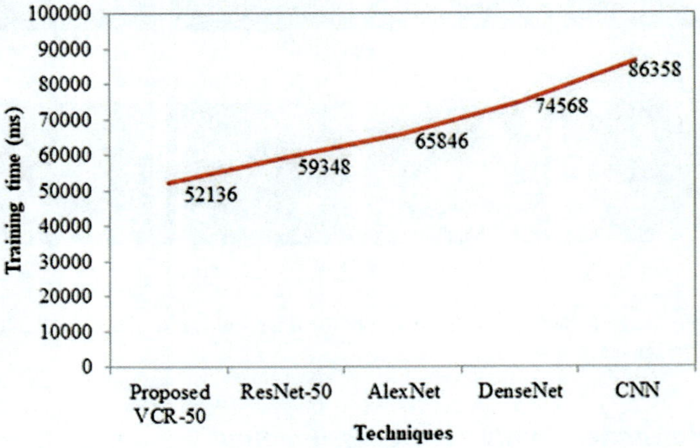


**Fig. 5** Training time of classifiers

## *4.3 Performance Analysis for Segmentation*

In this segment, the experimental results of proposed SVT-KLD-FCM segmentation are analyzed with traditional FCM, Fuzzy-K Means (FKM) and K-Means cluster-based segmentation techniques in terms of segmentation accuracy (SA), False Positive Rate (FPR) and False Negative Rate (FNR).

Table 1 depicts the segmentation accuracy of the proposed and prevailing segmentation techniques, in which the K-Means clustering-based segmentation algorithm gives a poor accuracy level of 89.73% and the SVT-KLD-FCM gains a higher SA level

**Table 1** SA outcomes

| Techniques | SA (%) |
|---|---|
| Proposed SVT-KLD-FCM | 96.27 |
| FCM | 94.05 |
| FKM | 91.62 |
| K-Means | 89.73 |

of 96.27%. This shows that the SVT-KLD-FCM is more suitable for the segmentation of BT.

## 4.4 *Performance Analysis of Edge Enhancement*

In this division, the experimental Peak Signal-to-Noise Ratio (PSNR) results of the proposed edge enhancement technique ACMPO are analyzed in comparison with PO, Sobel, Canny, and Laplacian of Gaussian (LoG) techniques.

In Fig. 6, it is proved that the PO attained a higher PSNR (25.3457 dB) value than the other traditional algorithms. But, with the ACM technique in PO, the PSNR increased by 2.495 dB, which shows the quality of the proposed edge-enhanced image.

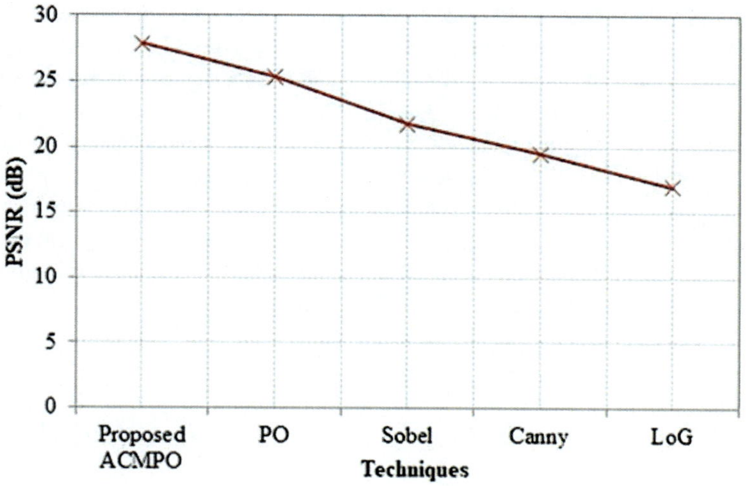


**Fig. 6** PSNR level of edge detection

**Table 2** Comparative analysis with the related works

| Techniques | AAccuracy (%) |
|---|---|
| Proposed VCR-50 | 98.77 |
| DCNN [17] | 95.72 |
| CNN-SVM [12] | 98.49 |
| LSTM [3] | 98 |

### *4.5 Comparative Analysis*

In this part, the proposed VCR-50 BT classifier is comparatively analyzed with the DCN [12, 17], and LSTM [3] models based on accuracy.

In Table 2, it is understood that the accuracy of the proposed VCR-50 (98.77%) BT classifier is higher than that of the DCNN, CNN-SVM, and LSTM models. This proves the dominance of the proposed BT classification over the other BT classification models.

## 5 Conclusion

This paper proposes BT classification with the VCR-50 classifier. In the proposed model, the tumor region is segmented with the SVT-KLD-FCM clustering-based segmentation algorithm. Also, the edge enhancement was performed with the ACMPO technique. The experiments were conducted on the BT MRI dataset. During experimental analysis, the proposed ACMPO achieved a higher PSNR level and the SVT-KLD-FCM attained an SA of 96.27%. Also, the proposed VCR-50 attained higher accuracy of 98.77%, recall of 98.35%, precision of 97.33%, and F-measure of 97.83% than the baseline classifiers with less training time of 52136 ms. Finally, the comparative analysis based on accuracy level proved the efficacy of the proposed classifier over other existing BT classification work. In this work, only three classes of BT are considered. Hence, in the future, along with the proposed classification, new classes of BT can be determined with the use of 3D images.

## References

1. Sharma P, Shukla AP (2021) A review on brain tumor segmentation and classification for MRI images. In: 2021 International conference on advance computing and innovative technologies in engineering, vol 7, pp 963–967
2. Shargunam S, Gopika RN (2020) Detection of brain tumor in medical imaging. In: 2020 6thInternational conference on advanced computing and communication systems, pp 416–420
3. Amin J, Muhammad S, Mudassar R, Tanzila S, Rafiq S, Shafqat AS (2020) Brain tumor detection: a long short-term memory (LSTM)-based learning model. Neural Comput Appl 32:15965–15973

4. Raut G, Raut A, Bhagade J, Bhagade J, Gavhane S (2020) Deep learning approach for brain tumor detection and segmentation. In: International conference on convergence to digital world-quo vadis, pp 1–25
5. Nazir M, Shakil S, Khurshid K (2021) Role of deep learning in brain tumor detection and classification (2015 to 2020): a review. Comput Med Imaging Graph 91:1–47
6. Siva Raja PM, Rani AV (2020) Brain tumor classification using a hybrid deep autoencoder with Bayesian fuzzy clustering-based segmentation approach. Biocybern Biomed Eng 40(1):440–453
7. Rehman A, Khan MA, Saba T, Mehmood Z, Tariq U, Ayesha N (2021) Microscopic brain tumor detection and classification using 3D CNN and feature selection architecture. Microsc Res Tech 84(1):133–149
8. Saba T, Sameh Mohamed A, El-Affendi M, Amin J, Sharif M (2020) Brain tumor detection using fusion of hand crafted and deep learning features. Cogn Syst Res 59:221–230
9. Hanif A, Doroslovacki M (2020) Graph laplacian-based tumor segmentation and denoising in brain magnetic resonance imaging, In:54th Asilomar conference on signals, systems, and computers, Pacific Grove, CA, USA, 2020, pp 241–245
10. Divyamary D, Gopika S, Pradeeba S, Bhuvaneswari M (2020) Brain tumor detection from MRI images using naive classifier. In: 6th international conference on advanced computing and communication systems (ICACCS), pp 620–622
11. Kumar DM, Satyanarayana D, Prasad MNG (2021) MRI brain tumor detection using optimal possibilistic fuzzy C-means clustering algorithm and adaptive k-nearest neighbor classifier. J Ambient Intell Hum Comput 12(2):2867–2880
12. Khairandish MO, Sharma M, Jain V, Chatterjee JM, Jhanjhi NZ (2022) A hybrid CNN-SVM threshold segmentation approach for tumor detection and classification of MRI brain images. IRBM 43(4):290–299
13. Tiwari P, et al. (2022) CNN based multiclass brain tumor detection using medical imaging. Comput Intell Neurosci 22:1–8
14. Ozbay E, Altunbey F (2023) Interpretable features fusion with precision MRI images deep hashing for brain tumor detection. Comput Methods Programs Biomed 231:1–13
15. Amin J, et al. (2019) SAC.: brain tumor detection by using stacked autoencoders in deep learning. J Med Syst 44(2):32
16. Sharif MI, Li JP, Amin J, Sharif A (2021) An improved framework for brain tumor analysis using MRI based on YOLOv2 and convolutional neural network. Complex Intell Syst 7(4):2023–2036
17. Ayadi W, Elhamzi W, Charfi I, Atri M (2021) Deep CNN for brain tumor classification. Neural Process Lett 53:671–700

# Abnormal Corner of Mouth Fall Detection of Stroke Patient Using Camera

**Piya Thirapanmethee, Jirayu Tancharoen, Khananat Sae-Tang, Nilubon Bootchai, Sirion Nutphadung, and Orasa Patsadu**

**Abstract** This paper proposes method to detect abnormal corner of mouth fall in stroke patient using camera. There are four steps of our proposed method. First, sample subject's face is captured. Next, face image is extracted for coordinate using image processing to obtain all coordinates. Then, appropriate coordinates are selected to compute degree of mouth corner (left and right). Finally, Multilayer Perceptron and Decision Tree are used to build model. From our experiment, the result shows that our proposed method for abnormality detection of corner of mouth fall in stroke patient using Decision Tree and Multilayer Perceptron achieve accuracy of 97.06% and 96.66%, respectively. From comparison of performance, Decision Tree achieves higher accuracy than Multilayer Perceptron. So, Decision Tree model is used to develop the prototype system for abnormality detection of corner of mouth fall in stroke patient. When the system detects abnormal corner of mouth fall in stroke patient, the system will send immediate notification to caregiver. Therefore, our proposed system could be the primary assistance, especially, in case the caregiver is not with the patient. So, caregiver can perceive symptoms and can make decisions in timely manner to refer a patient for further treatment.

**Keywords** Corner of Mouth · Stroke · Camera · Notification · Healthcare · Machine Learning

## 1 Introduction

In Thailand, "Stroke is the second leading cause of death after all types of cancer" [1]. The major cause of stroke is divided into three types: Ischemic stroke, Hemorrhagic stroke, and Transient Ischemic Attack [2]. "F.A.S.T is warning signal to indicate stroke" [3]. (F is Face: when the patient smiles or shows the teeth, the corner of the lips droops. A is Arm: the patient will not be able to raise up the arm(s). S is Speech:

P. Thirapanmethee · J. Tancharoen · K. Sae-Tang · N. Bootchai · S. Nutphadung · O. Patsadu (✉)
Computer Science, Faculty of Science and Technology, Rajamangala University of Technology Krungthep, Bangkok, Thailand
e-mail: orasa.p@mail.rmutk.ac.th



P. Meesad et al. (eds.), *Proceedings of the 19th International Conference on Computing and Information Technology (IC2IT 2023)*, Lecture Notes in Networks and Systems 679,
https://doi.org/10.1007/978-3-031-30474-3_3

the patient has a problem of slurred speech and has difficulty speaking. T is Time: the patient needs immediate treatment to reduce risk of disability and death, and to increase a chance to return to normal [3–5]. From the warning signals, one of the major notifying signals of stroke is an abnormal drooping of corner of lips. Reducing the chance of disability and death and performing self-assessment would prevent the severity of stroke. Therefore, the objective of our research is a real time system to monitor movement of patient for abnormality detection of corner of mouth fall in the early stage that achieves high accuracy and includes immediate notification to the caregiver, especially, for stay-at-home patients.

There are several research studies that proposed methods to detect stroke and facial paralysis, especially in case of abnormality detection of corner of mouth fall [6–14], which is abnormalities of cranial nerve seventh pair (Bell's palsy). The symptom of disease is an anesthesia or sudden bell's palsy, which leads to corner of mouth fall, crooked mouth, with or without excessive drooling from the corners of the mouth" [15]. However, the most of research studies lack movement monitoring system of patient to detect abnormal corner of mouth fall for immediate notification to caregiver. This gap becomes our research study.

This research study proposed a method for abnormality detection of corner of mouth fall in patients with risk of stroke using camera. The feature of our research study is the system where, when the patient exhibits symptom of abnormal corner of mouth fall, the system will send immediate notification to caregiver and refer patient to see a physician for further treatment. Therefore, our research study is divided into 2 phases: First, model of abnormality detection of corner of mouth fall and notification. For the first phase, our proposed method detects abnormal corner of mouth fall using face image. After that, face image is extracted for coordinates and the suitable coordinates are selected to compute degree of mouth corner fall (left and right). Finally, abnormality detection of corner of mouth fall uses a performance comparison of models such as Multilayer Perceptron and Decision Tree. In the second phase, if the system detects abnormal corner of mouth fall, the system will send immediate notification to caregiver, which will be helpful in case the caregiver is not with patient to give immediate help to reduce the chance of paralysis and death, the burden of a family, the cost of medication, and to increase chances of bodily recovery [16].

This remaining section of this paper is arranged as follows: related works section, methodology section, results and discussion section, and conclusion section.

## 2 Related Works

For stroke detection, several research studies use classification method [17–23]. Mekurai et al. [17] present a system to detect impaired balance of stroke using muscle weakness. The result shows that Artificial Neural Network achieve the accuracy of 93%. Thammaboosadee and Kansadub [18] detect stroke using demographic information and medical screening data via Naive Bayes, Decision Tree, and Artificial Neural Network. Chin et al. [19] propose stroke patient detection using CT scan image and Convolutional Neural Network, which yields accuracy higher than 90%. Dourado Jr et al. [20] classifies stroke using CT scan image and Convolutional Neural Network. The model uses Bayesian. The system detects stroke patient with the accuracy of 100%. Gaidhani et al. [21] propose stroke detection using MRI images. The features are extracted using Convolutional Neural Network and Deep Learning, which yield approximate accuracy of 96–97%. Chantamit-o-pas and Goyal [15] present a method to detect stroke using heart disease dataset and several classifiers such as Deep Learning, Naïve Bayes, and Support Vector Machine. Cheon et al. [22] detect stroke using medical information via Principal Component Analysis (PCA). Deep Neural Network is used to build model. The system can detect stroke with an Area under the Curve (AUC) of 83.48%. Li et al. [23] identify the risk level of stroke patients using several classifiers such as "Logistic Regression, Naïve Bayesian, Bayesian Network, Decision Tree, Neural Network, Random Forest, Bagged Decision Tree, and Voting by Boosting with Decision Trees" [23]. The result shows that Boosting with Decision Trees achieved recall of 99.94%.

For research studies related to detecting abnormal corner of mouth fall in stroke patient, Gupta [6] propose a system on smartphone to diagnose stroke patient using Deep Learning. The system records "heart rate, blood pressure, and blood oxygen of patient" [6]. In addition, smartphone captures face and records speech of patient to build model using Recurrent Neural Network, Support Vector Machine, and Convolutional Neural Network. The system can diagnose stroke patient with 95% accuracy. Foong et al. [7] detect abnormal corner of mouth fall using Google Mobile Vision using "National Cheng Kung University (NCKU) Robotics Face dataset" [8] to extract coordinates, which are used to compute degree of mouth corner. Park et al. [9] propose "real time detection system for stroke patient using air cushion seat and sensors during car drives of elderly. ECG, EEG, heart rate, seat pressure balance data, face or eye tracking are used to detect symptom of stroke" [9]. Chang et al. [10] detect abnormality of face of stroke patient. Face image is extracted to compute area ratio and distance ratio between the left and right side of the eye and mouth. The system achieved the accuracy of 100% using Support Vector Machine and Bayesian, and 95.45% using Random Forest. Umirzakova and Whangbo [11] detect stroke patient using face image, whose features consist of "wrinkles on forehead area, eye moving, mouth drooping, and cheek line detection" [11]. Parra-Dominguez et al. [12] propose facial paralysis detection of stroke patient using Face Landmark. Then model is built by Multilayer Perceptron, which yields accuracy of 94.06%. Parra-Dominguez et al. [13] estimate severity of palsy patients severity levels: healthy, slight, and strong palsy, with accuracy of 95.58%. Gaber et al. [14] use Kinect V.2, with facial paralysis detection using ensemble of Support Vector Machine. Severity levels are mild, moderate, and severe. The result yields accuracy of 96.8%. Other research studies proposed methods to detect symptom such as Bell's palsy or Facial Palsy, which occur because of weakened facial muscles or temporary paralysis on facial nerves. In Bell's palsy, "the patient cannot raise eyebrows, eyelid and corner of mouth fall, eyes not able to completely shut, and drooling from the corners of the mouth" [24]. Carro et al. [25] use Kinect to capture images that are later computed for coordinates from edges of mouth and lateral can thus. Guarin and Dusseldorp [26] propose a real time system to detect facial paralysis using face images, which is extracted for pupil position and center line of the face. The result is then used to compute deviation of face. Gaber et al. [27] detect facial paralysis using Kinect V.2 using facial symmetry. The results show that mean SI of eyebrows, mean SI of eyes and mean SI of mouth, each, detects with accuracy over 98%.

## 3 Methodology

In this section, we explain our method to detect abnormal corner of mouth fall in stroke patient using camera as shown in Fig. 1.

From Fig. 1, Model building for abnormality detection of corner of mouth fall in stroke patient is divided into 4 steps: data collection, data preparation, model building, and notification as described in next section.

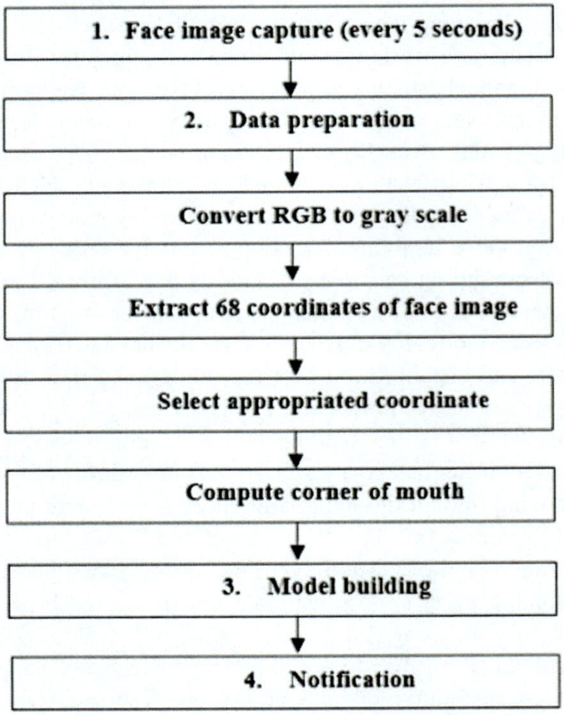


**Fig. 1** Steps for model building for abnormality detection of corner of mouth fall in stroke patient

## 3.1 Data Collection

In data collection, by approval of the Institutional Review Board (IRB), there are 10 sample subjects (5 stroke patients with persistent abnormality of corner of mouth fall and 5 normal sample subjects at the age of 55 $\pm$ 10). In our experiment, we set up camera approximately 50 cm. from sample the subject's face, which is a suitable distance. We capture approximately 11 frontal images of the subject's face in 5 s each time, for 20 times. There is a total of suitable 2,038 images (1,030 images of stroke patient with a persistent abnormal corner of mouth fall and 1,008 images of normal sample subjects).

## 3.2 Data Preparation

Once we obtained face image of sample subject, there is a total of 2,038 images as explained in Sect. 3.1. Then, face image of sample subject is extracted for features to build model as follows.

### 3.2.1 Image Conversion

The first step is an image conversion from RGB to grayscale using OpenCV [28].

### 3.2.2 Face Localization

In the second step, once we obtain grayscale image, image is extracted for coordinates using dlib [29], which obtain a total of 68 coordinates.

### 3.2.3 Suitable Coordinate Selection for Computation of Degree of Mouth Angle

For the third step, there is a total of 68 coordinates as obtained in Sect. 3.2.2. The principle of consideration for selecting suitable coordinate for model building is studied from the concept of Foong et al. [7] and Gaber et al. [27]. For research of Foong et al. [7], "there are 3 coordinates such as left mouth, right mouth, and bottom mouth for abnormality detection of corner of mouth fall on stroke patient". In the work of Gaber et al. [27], it detects facial paralysis and evaluate facial symmetry using 3 coordinates such as eyebrows, eyes, and mouth to compute area of eyes and slope of mouth. We have found that suitable coordinates for abnormality detection of corner of mouth fall in stroke patient consist of 4 coordinates (27, 57, 48, and 54)

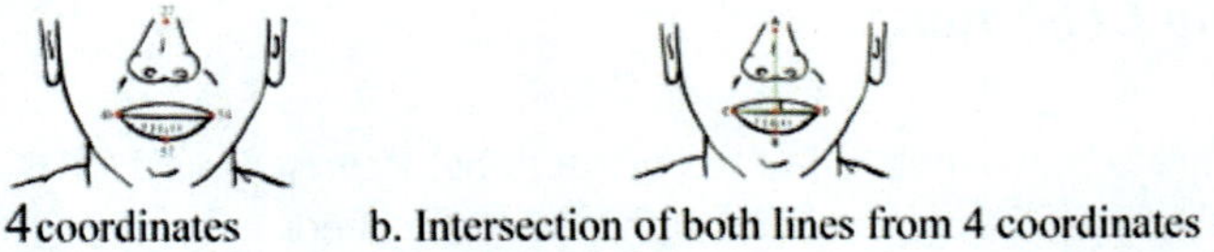


**Fig. 2** Suitable coordinate selection

**Table 1** Comparative result of accuracy of suitable coordinate selection for abnormality detection of corner of mouth fall in stroke patient

| Coordinates on face | Accuracy of detection |
|---|---|
| 3 coordinates (left mouth, right mouth, and bottom mouth) [7] | 93.40% |
| 3 coordinates (eyebrows, eyes, and mouth) and computation of slope of mouth [27] | 95.65% |
| Our proposed 4 coordinates (the nose bridge, a middle of the lower lip, left corner of the mouth, and right corner of the mouth) and computation of corner of mouth | 97.06%* |

* High accuracy

as seen in Fig. 2(a), whose result of accuracy is better than using other coordinates as seen in Table 1.

From Table 1, the result shows suitable coordinate selection for abnormality detection of corner of mouth fall in stroke patient in several cases. The result shows that 3 coordinates (left mouth, right mouth, and bottom mouth) [7] achieves accuracy of 93.40%. In addition, the 3 coordinates (eyebrows, eyes, and mouth) and computation of slope of mouth [27] achieves accuracy of 95.65%. However, to enhance effectiveness, we should use 4 coordinates (the nose bridge, a middle of the lower lip, left corner of the mouth, and right corner of the mouth) and uses these coordinates to compute corner of mouth. From our experiment, the result shows that the 4 coordinates achieve accuracy of 97.06%. We then select the 4 coordinates for the detection.

### 3.2.4 Computation of Corner of Mouth

From our experiment to select suitable coordinate for computation of corner of mouth fall in stroke patient in the initial stage. We use 4 coordinates to compute corner of mouth as explained in Sect. 3.2.3. For the first coordinate (27), it is the nose bridge (position A). The second coordinate (57) is a middle of the lower lip (position B). The remaining two coordinates (48 and 54) are left corner of the mouth (position C) and right corner of the mouth (position D) as shown in Fig. 2(a), (b). Then, these coordinates are used to compute the intersection of both lines to be used to compute corner of mouth as shown in Fig. 2(b).

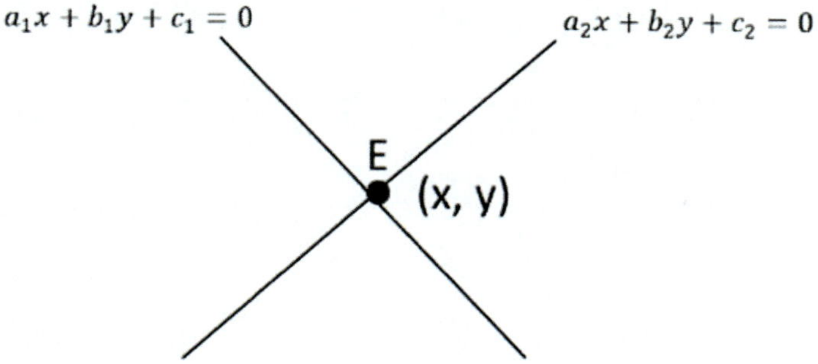


**Fig. 3** Intersection of both lines [30]

From Fig. 2(b), we compute the degree of angle of left mouth corner and right mouth corner from the intersection of both lines (position E) to divide left and right-hand side of face. Then, we obtain (x, y) coordinate from the intersection of both lines as shown in Fig. 3 [30].

Once, we obtain (x, y) coordinate from the intersection of both lines position E, the (x, y) coordinate is used to compute angle (in Radian) using arctan2 function [31]. Then, angle of Radian is used to convert to degree [32] as shown in Eq. 1.

$$degree = \frac{(radian * 180)}{\pi} \tag{1}$$

Radian is a unit of angle.
$\pi$ *is* a mathematical constant (3.14159).

### *3.3 Model Building for Abnormality Detection of Corner of Mouth Fall in Stroke Patient*

Relying on data preparation as explained in Sect. 3.2, we use a dataset with 2,038 rows (1,030 rows of stroke patient with a persistent abnormal corner of mouth fall and 1,008 rows of normal sample subjects), with each consists of 3 attributes namely degree of left mouth corner, degree of right mouth corner, and class label (stroke patient with a persistent abnormal corner of mouth fall or normal sample subject). For model building, we compare performance of 2 classifiers namely Decision Tree [33, 34] and Multilayer Perceptron [34, 35]. The model is trained and tested with tenfold cross validation [34] to reduce chance of overfitting [34]. Each parameter is identified as follows:

Decision Tree [33, 34] is a model that is not the complex and is an easy classification. Decision Tree is easy to understand and can be used to solve problem of classification. In our experiment, we set parameters of Decision Tree with a confidence value (0.5) and seed (1).

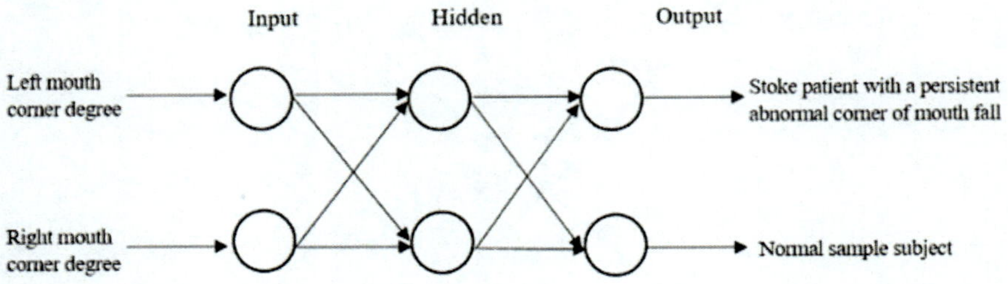


**Fig. 4** Model for abnormality detection of corner of mouth fall in stroke patient using Multilayer Perceptron

Artificial Neural Network based on Multilayer Perceptron [34, 35] is classification method, which is highly flexible and robust to dataset of error correction, because it has parameter tuning to enhance the performance of model. We set parameter of Multilayer Perceptron as follow: input node (2 values: left mouth corner degree and right mouth corner degree), 2 hidden nodes, 0.3 learning rate, 0.2 momentum, and output node (2 values: stroke patient with a persistent abnormal corner of mouth fall and normal sample subject) as shown in Fig. 4.

### 3.4 *Notification*

From the obtained result in Sect. 3.3, if the system detects abnormal corner of mouth fall, the system will send immediate notification to the caregiver via smartphone message [36]. Therefore, the caregiver can acknowledge the situation and make needed decisions in timely manner. In addition, the caregiver and family can communicate via message to coordinate assistance and to refer patient to hospital for further treatment.

## 4 Results and Discussion

There are two main points, which is explained in this section (the result of performance measurement of model and case study of our testing of the proposed system) as follows.

## 4.1 Result of Performance Measurement of Model

We explain the result of performance measurement of the model by comparing of 2 classifiers namely Multilayer Perceptron (MLP) and Decision Tree (DT) as seen in Table 2.

From Table 2, the result of abnormality detection of corner of mouth fall in stroke patient using Decision Tree achieves a high accuracy of 97.06%. Multilayer Perceptron can detect abnormal corner of mouth fall in stroke patient with an accuracy of 96.66%. The reason why of Decision Tree can detect abnormal corner of mouth fall in stroke patient with higher accuracy than Multilayer Perceptron is because it can classify answers, which is independent to data distribution. In addition, Decision Tree can detect accuracy in several cases, for example, the case of sample subject smiling and speaking that slightly resemble the case of corner of mouth fall.

From performance comparison of the two models, the result shows that Decision Tree achieves higher performance than Multilayer Perceptron. Therefore, we select Decision Tree as the model to develop a prototype system, which is an intervention program for abnormality detection of corners of mouth fall in stroke patient.

In addition, to reduce biased data in our proposed method, we use our dataset and dataset of "National Cheng Kung University (NCKU) Robotics Face database" [8] combined to detect abnormal corner of mouth fall in stroke patient as seen in Table 3.

From Table 3, the results show abnormality detection of corner of mouth fall in stroke patient using our dataset (stroke patient with a persistent abnormal corner of mouth fall) and dataset of "National Cheng Kung University Robotics Face database" (normal sample subjects) [8]. There are 90 sample subjects, which make up 6,660 images (frontal image ($0^0$) and tilted face image (left and right). However, we use 179 frontal images ($0^0$) to build the model. The result shows that the system can detect abnormal corner of mouth fall in stroke patient using Decision Tree and Multilayer Perceptron with an accuracy of 98.01% and 97.93%, respectively.

Moreover, we use our dataset (stroke patient with a persistent abnormal corner of mouth fall) and dataset of "National Cheng Kung University Robotics Face database (normal sample subjects of the frontal image ($0^0$) and tilted face images (left and right with $\pm 5^0$ and $\pm 10^0$) [8]. There is a total of 895 images. The result shows that Decision Tree can detect abnormal corner of mouth fall in stroke patient with an

**Table 2** Result of performance comparison of multilayer perceptron and decision tree

| | Multilayer perceptron | Decision tree |
|---|---|---|
| Accuracy | 96.66% | 97.06%* |
| Precision | 96.70% | 97.20% |
| Recall | 96.70% | 97.10% |
| F-measure | 96.70% | 97.10% |

* High accuracy

**Table 3** Result of compared performance of detection

| | Our proposed method on our dataset | | Our proposed method (our dataset and dataset of "National Cheng Kung University Robotics Face database" only fatal face image [8] | | Our proposed method (our dataset and dataset of "National Cheng Kung University Robotics Face database" on several views such as $0^0$, $\pm 5^0$, $\pm 10^0$) [8] | |
|---|---|---|---|---|---|---|
| | MLP DT | DT | MLP | DT | MLP | DT |
| Accuracy | 96.66% | 97.06%* | 97.93% | 98.01%* | 93.82% | 94.13%* |
| Precision | 96.70% | 97.20% | 97.90% | 98.00% | 94.20% | 94.30% |
| Recall | 96.70% | 97.10% | 97.90% | 98.00% | 93.80% | 94.10% |
| F-measure | 96.70% | 97.10% | 97.90% | 98.00% | 93.80% | 94.10% |

* High accuracy

accuracy of 94.13% and Multilayer Perceptron can detect abnormal with an accuracy of 93.82%.

As we use another dataset for abnormality detection of corner of mouth fall in stroke patient to reduce biased dataset in our proposed method, it can be concluded that Decision Tree achieves higher accuracy than Multilayer Perceptron overall. Therefore, it is guaranteed that Decision Tree model can be used in the application of the model.

## 4.2 *Case Study*

In the case study of abnormality detection of corner of mouth fall in stroke patient, suppose that the patient sits down and then the patient has a symptom of corner of mouth fall as shown in Fig. 5.

From Fig. 5, when the system detects abnormal corner of mouth fall in stroke patient, the system will send immediate notification to caregiver to report symptom of patient via smartphone. The notification consists of message ("detected abnormal corner of mouth fall"), timestamp, and image during occurred situation.

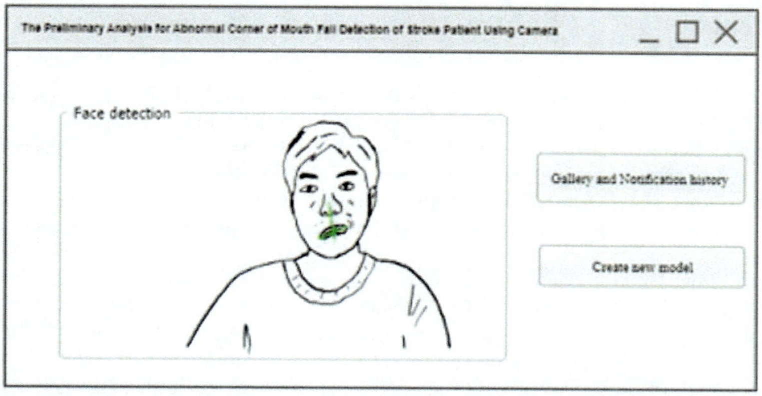


**Fig. 5** Case study

# 5 Conclusions

This paper proposed method to detect abnormal corner of mouth fall in stroke patient using a camera. The first step is face image capture of sample subjects. In the second step, the face image is extracted for coordinates using image processing. The third step is the selection of appropriate coordinate such as the bridge of the nose point, the middle of the lower lip point, and the corners of the mouth point (left and right) to compute degree of mouth corner (left and right) for model building. Multilayer perceptron and Decision Tree are used to build model for abnormality detection of corner of mouth fall in stroke patient. From our experiment, the result shows that our proposed method can detect abnormal corner of mouth fall in stroke patient using Multilayer perceptron and Decision Tree with an accuracy of 96.66% and 97.06%, respectively. So, Decision Tree achieves higher accuracy than Multilayer perceptron, it is used to develop the prototype system. Moreover, we use our dataset (stroke patient with a persistent abnormal corner of mouth fall) and dataset of "National Cheng Kung University robotics face database (normal sample subjects)" [8] to detect abnormal corner of mouth fall in stroke patient. The result shows that the system can detect abnormal corner of mouth fall in stroke patient using Decision Tree and Multilayer Perceptron with an accuracy of 98.01% and 97.93%, respectively. In addition, Decision Tree can detect abnormal corner of mouth fall in stroke patient with an accuracy of 94.13% and Multilayer Perceptron can detect with accuracy of 93.82% in several views ($0^0$, $\pm 5^0$, and $\pm 10^0$). Finally, when the system detects abnormal corner of mouth fall in stroke patient, the system will send immediate notification to caregiver. The caregivers can perceive symptoms of the patient remotely. Moreover, the caregiver can make needed decision in timely manner and refer patient to hospital to reduce the chance of paralysis and to increase chance of recovery. Furthermore, our proposed system can extend to be a smart home care to save life of patients.

For future work, we will develop the system to detect stroke by the integration of other factors such as imbalance gait to improve performance of our proposed system.

## References


1. Office of Policy and Strategy of the Ministry of Public Health (2018) Public Health Statistics A.D.2018. WVO office of Printing Mill, Bangkok, ISSN 0857-3093
2. Wei CC, Huang SW, Hsu SL, Chen HC, Chen JS, Liang H (2012) Analysis of using the tongue deviation angle as a warning sign of a stroke. Biomed Eng 11(53):1–12
3. World stroke Day, 2018. FAST. https://www.timesnownews.com/health/article/world-stroke-day-2018-theme-fast-10-warning-signs-and-symptoms-of-stroke/305917. Accessed 21 Nov 2022
4. World stroke Organization. https://www.world-stroke.org/. Accessed 21 Nov 2022
5. American stroke Association, stroke Warning Signs and Symptoms. http://www.strokeassociation.org/STROKEORG/WarningSigns/strokeWarning-Signs-and-Symptoms_UCM_308528_SubHomePage.jsp. Accessed 15 Sep 2022
6. Gupta A. (2019) strokeSave: a novel, high-performance mobile application for stroke diagnosis using deep learning and computer vision. ArXiv 1–16
7. Foong OM, Hong KW, Yong SP (2016) Droopy mouth detection model in stroke warning. In: 3rd International conference on computer and information sciences, pp 616–621, Kuala Lumpur, Malaysia
8. National Cheng Kung University (NCKU) Robotics Face Dataset, Databases for Face Detection and Pose Estimation. http://robotics.csie.ncku.edu.tw/Databases/FaceDetect_PoseEstimate.htm#Our_Database. Accessed 9 Sep 2022
9. Park S, Hong S, Kim D, Seo Y (2018) Development of a real-time stroke detection system for elderly drivers using Quad-Chamber air cushion and IoT devices. SAE Technical 1–5
10. Chang CY, Cheng MJ, Ma MHM (2018) Application of machine learning for facial stroke detection. In: 23rd international conference on digital signal processing, pp 1–5, Shanghai, China
11. Umirzakova S, Whangbo TK (2018) Study on detect stroke symptoms using face features. In: International conference on information and communication technology convergence, pp 429–431, Jeju, Korea (South)
12. Parra-Dominguez GS, Sanchez-Yanez RE, Garcia-Capulin CH (2021) Facial paralysis detection on images using key point analysis. Appl Sci 21(2435):1–11
13. Parra-Dominguez GS, Garcia-Capulin CH, Sanchez-Yanez RE (2022) Automatic facial palsy diagnosis as a classification problem using regional information extracted from a photograph. Diagnostics 12(1528):1–17
14. Gaber A, Taher MF, Wahed MA, Shalaby NM, Gaber S (2022) Classification of facial paralysis based on machine learning techniques. Biomed Eng Online 21(65):1–20
15. Chantamit-o-pas P, Goyal M (2017) Prediction of stroke disease using deep learning model. Neural Inf Process 10638:1–9
16. Suwiwattana S, Kasemset C, Khwanngern K (2020) Healthcare service network analysis: Northern region's healthcare service network of cleft lip and cleft palate. Curr Appl Sci Technol 20(2):198–207
17. Mekurai C, Rueangsirarak W, Kaewkaen K, Uttama S, Chaisricharoen R (2020) Impaired balance assessment in older people with muscle weakness caused by stroke. ECTI Trans Comput Inf Technol 14(2):103–112
18. Thammaboosadee S, Kansadub T (2019) Data mining model and application for stroke prediction: a combination of demographic and medical screening data approach. J Thai Interdiscipl Res 14(4):61–69
19. Chin CL, Lin BJ, Wu GR et al (2017) An automated early Ischemic stroke detection system using CNN deep learning algorithm. In: 8th International conference on awareness science and technology, pp 368–372, Taiwan
20. Dourado Jr CMJM, da Silva SPP, da Nóbrega RVM, Barros ACDS, Filho PPR, de Albuquerque VHC (2019) Deep learning IoT system for online stroke detection in skull computed tomography images. Comput Netw 152:25–39

21. Gaidhani BR, Rajamenakshi R, Sonavane S (2019) Brain stroke detection using convolutional neural network and Deep Learning models. In: 2nd International conference on intelligent communication and computational techniques, pp 242–249, Jaipur, India
22. Cheon S, Kim J, Lim J (2019) The use of deep learning to predict stroke patient mortality. Int J Environ Res Public Health 16(1876):1–12
23. Li X, Bian D, Yu J, Li M, Zhao D (2019) Using machine learning models to improve stroke risk level classification methods of China national stroke screening. BMC Med Inform Decis Mak 19(261):1–7
24. Seta DD, Mancini P, Minni A et al (2014) Bell's palsy: symptoms preceding and accompanying the facial paresis. Sci World J 2014:1–6
25. Carro RC, Huerta EB, Caporal RM, Hernández JC, Cruz FR (2016) Facial expression analysis with Kinect for the diagnosis of paralysis using Nottingham system. IEEE Lat Am Trans 14(7):3418–3426
26. Guarin DL, Dusseldorp J (2018) A machine learning approach for automated facial measurements in facial palsy. JAMA Facial Plast Surg 20(4):1–3
27. Gaber A, Taher MF, Wahed MA (2015) Quantifying facial paralysis using the Kinect v2. In: Annual international conference of the IEEE engineering in medicine and biology society, pp 2497–2501, Milan, Italy
28. Opencv https://opencv.org. Accessed 1 Sep 2022
29. King D (2022) Dlib C++ library. http://dlib.net. Accessed 2 Sep 2022
30. Point of Intersection Formula. https://byjus.com/point-of-intersection-formula. Accessed 10 Sep 2022
31. MedCalc Software. ATAN2 function. https://www.medcalc.org/manual/atan2_function.php. Accessed 18 Sep 2022
32. Mathcentre, Radians. http://www.mathcentre.ac.uk/resources/workbooks/mathcentre/mc-TY-radians-2009-1.pdf. Accessed 18 Sep 2022
33. Badulescu LA (2006) Data Mining Algorithms Based on Decision Trees. Annals of the Oradea University, Fascicle of Management and Technological Engineering, pp 1621–1628
34. Han J, Kamber M (2011) Data mining concepts and techniques, 3rd edn, Morgan Kaufmann Publishers
35. Popescu MC, Balas VE, Popescu LP, Mastorakis N (2009) Multilayer perceptron and neural networks. WSEAS Trans Circ Syst 8(7):579–588
36. Line application. https://notify-bot.line.me/doc/en/. Accessed 26 Sep 2022

# Federated Machine Learning for Self-driving Car and Minimizing Data Heterogeneity Effect

**Prastav Pokharel and Babu R. Dawadi**

**Abstract** This study has implemented federated learning concept to train the car to make it autonomous. Data is collected with the help of simulated car developed by udacity for two different tracks. It records images from center, left, and right cameras with associated steering angle, speed, throttle, and brake. Then using Convolution Neural Network, it is trained to form the modal. After training, the modal is submitted to the server where the models from two different sources that are combined together to generate new modal which is further sent to client for further training. Multiple training have been carried out to analyze the performance of car in autonomous mode. We found that accuracy is not always dependent upon number of iteration. Also, the combined model has always less accuracy than individual model for that specific track from where it is generated. The server initializes the model and global control variate (c) and pass it to the entire client i.e. car for our case. After receiving initial model and control variate, the car will update model and its local control variate (ci). With the help of correction term (c – ci), the server will converge the model in right direction minimizing the effect of data heterogeneity or client drift. Along with implementing federated machine learning, we focus on minimizing the effect of data heterogeneity that arises while training.

**Keywords** autonomous driving · machine learning · federated learning · data heterogeneity

## 1 Introduction

Decentralized artificial intelligence refers to shifting learning process to local devices so that they could make their own decision. Individualized decentralized learning has the disadvantage that, unlike centralized learning, results from learning obtained on other devices cannot be generalized. Combining centralized and personalized

P. Pokharel · B. R. Dawadi (✉)
Department of Electronics and Computer Engineering, IOE Pulchowk Campus, Tribhuvan University, Kathmandu, Nepal
e-mail: baburd@ioe.edu.np



P. Meesad et al. (eds.), *Proceedings of the 19th International Conference on Computing and Information Technology (IC2IT 2023)*, Lecture Notes in Networks and Systems 679,
https://doi.org/10.1007/978-3-031-30474-3_4

decentralized learning is a challenging and involved topic that calls for many machine learning (ML) approaches to be used. A new research field, Federated Learning (FL), is addressing this issue. In essence, FL is an ML technique to train algorithms across decentralized edge devices while holding data samples locally.

To make the car fully autonomous, it has to be trained in different location, climate, etc. which is not feasible. If we use federated learning concept, the knowledge obtained by one car after training could be used by another car since, we have the common server for communication. The main problem while making self-driving car are: (a) several variations in road structure and uncertain traffic situation, (b) training in one place will not be suitable for other geographically diverse places, (c) for central server, to train all the data coming from almost all parts of the world is highly costly, and (d) data generated by car from different location will be heterogeneous that causes model biasing problem for specific data set.

The main application of this study will be in self-driving car. At a large scale, all the vehicles from most part of the world will train in parallel with the data obtained from their environment. In the meantime, we can have trained model from all over the world in different weather, different geographical situation, and different uncertainty to improve the performance of self-driving car.

The rest of the paper is organized as follows. Section 2 provides the related works. Section 3 presents the detail methodology of environment setup, experimental evaluation, simulation, and analysis. Section 4 is dedicated to detail implementation including results and analysis of our work, while Sect. 5 concludes the paper.

## 2 Related Work

Autonomous driving is considered as the big disruptive innovation in the future. However, the level 5 cars' complete automation necessitates mastery of the many issues that surround their creation and entry to the market, including as the identification of other road users and the observation of driver behavior in case manual control needs to be restored [1]. With the idea of self-driving automobiles, the automotive business is rapidly changing, and all the corporations are concentrating on creating their own autonomous vehicles. Even businesses like Google and Uber, which are not involved in the "mainstream automotive," are heavily investing in and studying autonomous vehicles. Apple is also advancing its "Titan" autonomous vehicle project. The idea of electric vehicles is already being used in real life. Both General Motors and Tesla have successfully introduced their respective electric vehicles to the market and are now accessible to consumers. Although research on autonomous vehicles is still ongoing, certain autonomous vehicles are now on the market, such as Tesla autopilot and GM super cruise control. The self-driving technology startup Waymo, a division of Google's parent company, is successfully testing its concept. Waymo has also stated that it will introduce autonomous trucks for delivery of products. Using Waymo open datasets, the approach to learn long short term memory based model is developed [2].

Complex mechanisms are required to safely transport people or freight in driverless cars from one place to another. With the introduction of AI-based AVs, many difficulties are experienced when they are used on public roads. Given the existing formalism and explainability of neural networks, it is impossible to demonstrate the functional safety of these vehicles, which poses a significant issue. Overcoming these difficulties has been made possible in large part by deep learning and AI [3]. Reinforcement learning could be the best way to train car for making it autonomous but it involves non affordable trial-and-error in real environment training. So the car is first trained in virtual environment and then transfer into real environment [4]. Since the reinforcement learning methods does not work well on complex urban scenarios, a framework to enable model-free deep reinforcement learning is developed where specific input representation is designed and visual encoding is used to capture the low-dimensional latent states [5]. Additionally, since driving includes complicated interactions between numerous intelligent agents in a highly non-stationary environment, it is suggested that linked autonomous driving problems be formulated using partially observable markov games (POSG) [6]. In straightforward driving situations that need for short prediction horizons, conventional behavior prediction algorithms are applicable. Deep learning-based approaches have gained popularity most lately due to their promising performance in contexts that are more complicated than those addressed by conventional approaches. For predicting vehicle behavior, various deep learning-based methods are applied [7].

There are several FL methods such as FedAvg [8], FedOpt [9], and FedMA [10]. These methods face problems in training the large CNNs. For large CNNs, recent works on FedNAS [11, 12] is ongoing, but they also rely on GPU training for completing the evaluations. Other methods [13]optimizes the communication cost but they don't consider edge computational capabilities. Although model parallelism-based split learning [14, 15] tries to break the constraint of computatuin, it requires frequent communication with the server. Knowledge Distillation [16] is used in a different manner from concurrent works [17]. Transferring knowledge from a large network to a smaller one [18] is only considered in previous work. But, all teachers and students here share the same dataset [19], while in our setting, each member (client) can only access its own independent dataset. Efficient On-device Deep Learning such as model compression [20], MobileNets [21], ShuffeNets [22], SqueezeNets [23], EfficientNets [24] are also been used. But, all of these techniques were used in the inference phase rather than the training phase.

## 3 Methodology

### *3.1 System Model*

Our proposed system model is shown in Fig. 1. It is divided into two parts that are client and server. Server initializes model and send it to the clients. Clients update the

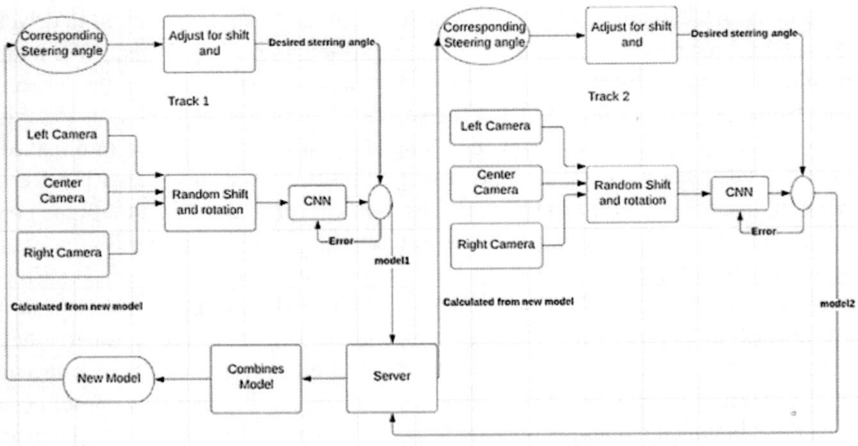


**Fig. 1** System model

model provided by the server by training in their specific tracks locally as mentioned above. Clients send the updated model to the server and then server aggregates it. After aggregation, server sends the updated model back to the clients. This process will continue for a specific number of times.

## 3.2 *Training Locally*

**Data Generation:**
Since the main objective of this thesis work is to implement federated machine learning concept for the self-driving car and to find out how the model biasing problem for certain type of environment could be solved we have used data which were generated from simulation. We have generated data from four similar tracks and one different track to analyze model biasing scenario.

Self-driving car simulator, which is open sourced by udacity is used for data generation purpose. It generates csv file to the desired location in PC. Keyboard is used to run the simulated car for data collection. The data collected have following properties, (i) image from center camera, (ii) image from left camera, (iii) image from right camera, (iv) steering angle, (v) throttle value, (vi) reverse status, (vii) speed, and (vii) brake status. The data generated by simulator is shown in Fig. 2.

**Data Preprocessing:**
The data collected was stored in a csv file. Following steps are done for data preprocessing, (i) Cropping the image for removing sky at the top and car front at the bottom, (ii) Resizing the image into input shape that is used by network model, (iii) Image is converted to YUV from RGB,(iv) Randomly choosing an image from center, left or right and adjust the steering angle, (v) Randomly flipping the image left

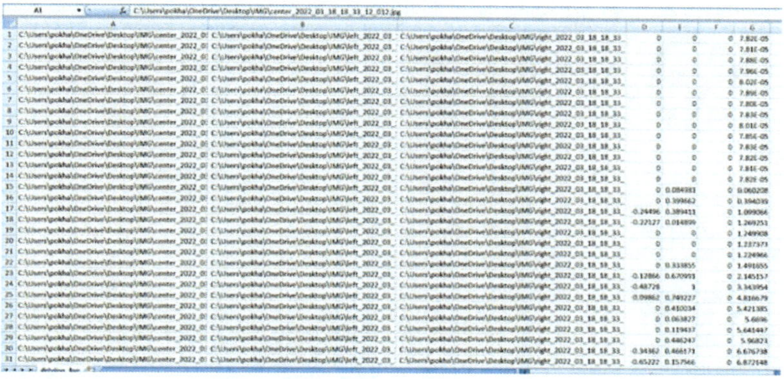

**Fig. 2** Collected data samples

<-> right, and adjust the steering angle, (vi) Randomly shifting the image vertically and horizontally (translation), (vii) Randomly adjusting brightness of the image, (viii) Generating an augmented image and adjust steering angle, and (ix) Generating training image which gives image paths and associated steering angles.

**Training Mode:**
Nine layers CNN is used. Image captured from the simulated car is provided to CNN layers, which contains three $5 \times 5$ convolutional layer using $2 \times 2$ strides and several filter values like (24, 36, 48, and 64). It will also contain drop out layer to prevent overfitting. It also contains fully connected neurons. We will use exponential linear unit for activation function. The output of this CNN will be steering value, throttle value, reverse status, speed, and brake status. After training the model provided by server initially is updated and also local control variate is updated.

**Model Aggregation:**
Server initializes the initial model and global control variate. The global control variate is used to identify how much the model updated by client deviate from global objective. These initialized parameters will be sent to clients i.e. car for update. Server receives the updated parameters from available clients. It aggregates the updates as well as control variate. For model aggregation server uses Federated average and scaffold algorithm (Fig. 3).

## 4 Implementations, Results and Analysis

The major steps carried out during the implementation of system are: (a) create the server, (b) server initialize the initial model and global control variate, (c) initial model and control variate are sent to clients (simulated car in different tracks in our case), (d) clients update model and control variate from local training environment

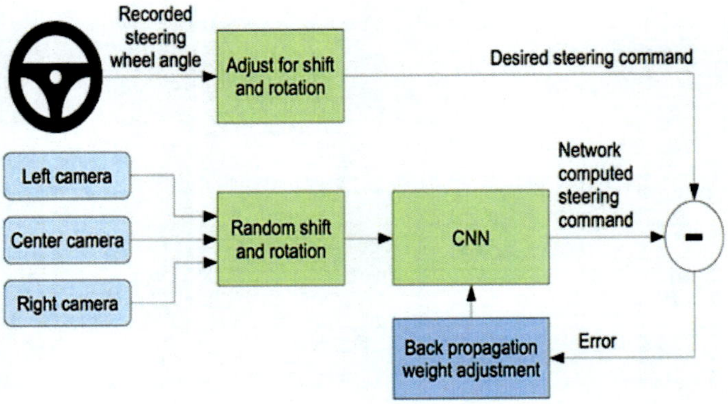


**Fig. 3** Training process block diagram

(data generated by simulator in given tracks), (e) updated model and control variate are re-send to server, (f) server generates the aggregated model by using federated average and scaffold algorithm and updates control variate, (g) the aggregated model and control variate are re-sent to clients, (h) this process is repeated ten times, (i) the accuracy of the final model obtained using federated average with scaffold algorithm are compared and analyzedto see reduced effecton biasing problem.

First of all, data is collected by driving car to different tracks in udacity car simulator on training mode. Here five tracks are used for training process. Out of which four tracks contains similar environment but fifth track have different environment i.e. different road structure. The images of first and fifth tracks are given in Fig. 4 and Fig. 5. Since other tracks have similar structure as first tracks they are not included in the figures.

Car was driven for few laps. Corresponding value of steering angle, reverse, throttle and speed along with image were collected by three different cameras.

After data is preprocessed, the model is designed for training those data. Image normalization is done to avoid saturation and make gradients work better. The model structure is shown in Fig. 6.

**Fig. 4** Data collection from track 1

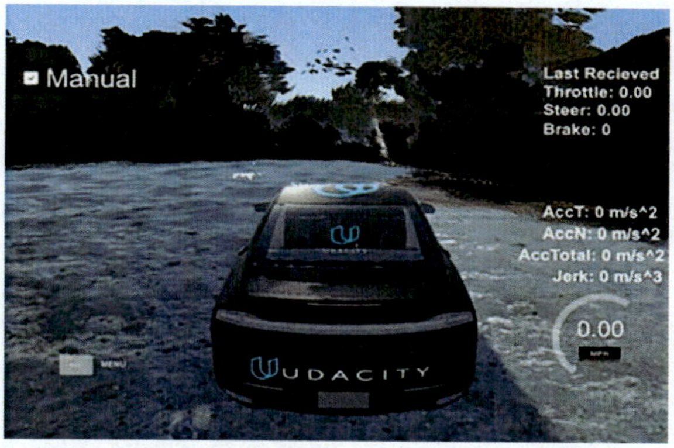


**Fig. 5** Data Collection from track 5

**Fig. 6** Model structure

| Layer (type) | Output Shape | Param # |
|---|---|---|
| lambda (Lambda) | (None, 66, 200, 3) | 0 |
| conv2d (Conv2D) | (None, 31, 98, 24) | 1824 |
| conv2d_1 (Conv2D) | (None, 14, 47, 36) | 21636 |
| conv2d_2 (Conv2D) | (None, 5, 22, 48) | 43248 |
| conv2d_3 (Conv2D) | (None, 3, 20, 64) | 27712 |
| conv2d_4 (Conv2D) | (None, 1, 18, 64) | 36928 |
| dropout (Dropout) | (None, 1, 18, 64) | 0 |
| flatten (Flatten) | (None, 1152) | 0 |
| dense (Dense) | (None, 100) | 115300 |
| dense_1 (Dense) | (None, 50) | 5050 |
| dense_2 (Dense) | (None, 10) | 510 |
| dense_3 (Dense) | (None, 1) | 11 |

The convolution layers are meant to handle feature engineering the fully connected layer for predicting the steering angle. Drop out layer is used to prevent overfitting. Similarly, exponential linear unit function takes care of the vanishing gradient. Following parameters are used for the training purpose: (i) Test size fraction: 0.2, (ii) Drop out probability: 0.5, (iii) Number of Epochs: 10, (iv) Samples per epochs: 20,000, (v) Batch Size: 40, an (vi) Learning Rate: 1.0e−4.

Model is generated after training the car. Specifically, the model generates the control value for car as shown in Fig. 7.

The simulator passes the images to our program. On the basis of final model provided by server, the steering angle, throttle, speed and brake is calculated. These

**Fig. 7** Calculated control values

```
(14516) wsgi starting up on http://0.0.0.0:4567
connect  d48ff022e6854e7aa7bf0b48e4d1e614
(14516) accepted ('127.0.0.1', 52930)
-0.24592038989067078 0.9395231618200205 0.0001
-0.24538934230804443 0.9397840706656254 0.0001
-0.24503517150878906 0.9399577647076583 0.0001
-0.24490603804588318 0.9399547729286685 0.2035
-0.21469628810882568 0.9477545526882921 1.9607
-0.2233089804649353 0.9250754756277112 3.9574
-0.2591654062271118 0.8727435947911361 6.1283
-0.19983908534049988 0.8617946295702724 7.837
-0.16202056407928467 0.8348052833114304 9.3188
```

calculated values are sent to simulated car from our so that it will move autonomously and further sent the next images to our program and this process continues. Socket input/output communication has been established, initially so that the information flow is easily managed form simulated car to the system.

In this way, client car updates the model by training in their own tracks. The updated model is sent to the server and server re-send aggregated model generated from the model sent by cars trained in other different tracks. This process is repeated ten times. Using federated average algorithm, the accuracy per epoch for fifth track is shown in Fig. 8.

Also using scaffold algorithm, the final accuracy per epoch for each trackis shown in Figs. 9, 10, 11, 12, and 13.

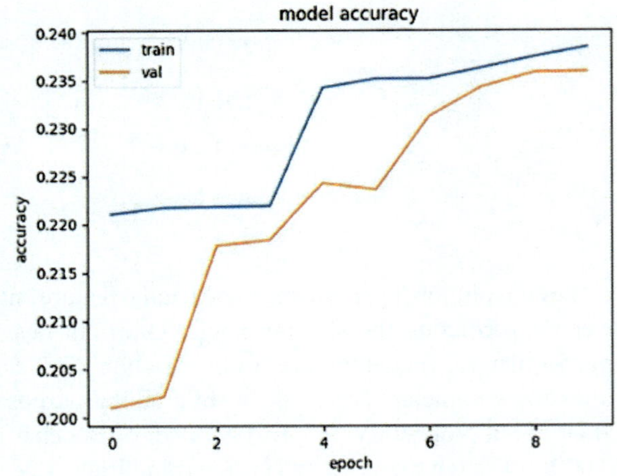


**Fig. 8** Accuracy per epoch after 10 client server update for track 5 for FedAvg

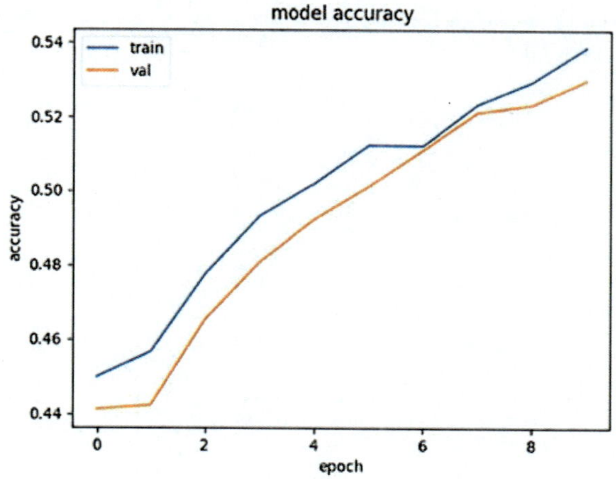


**Fig. 9** Accuracy per epoch for track 1

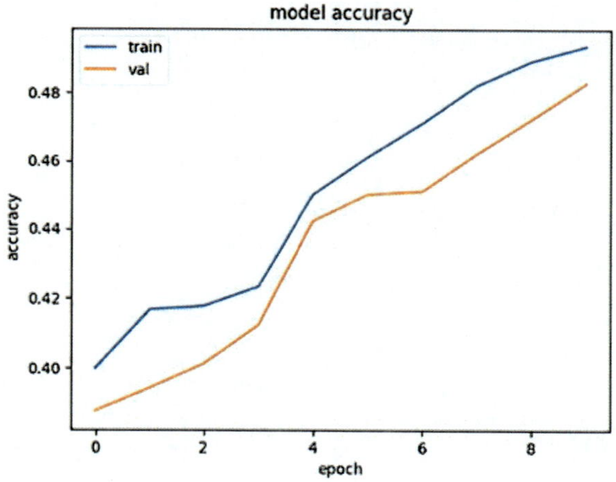


**Fig. 10** Accuracy per epoch for track 2

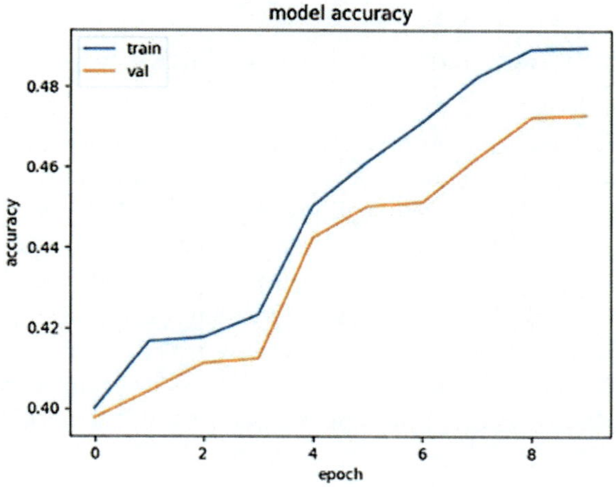


**Fig. 11** Accuracy per epoch for track 3

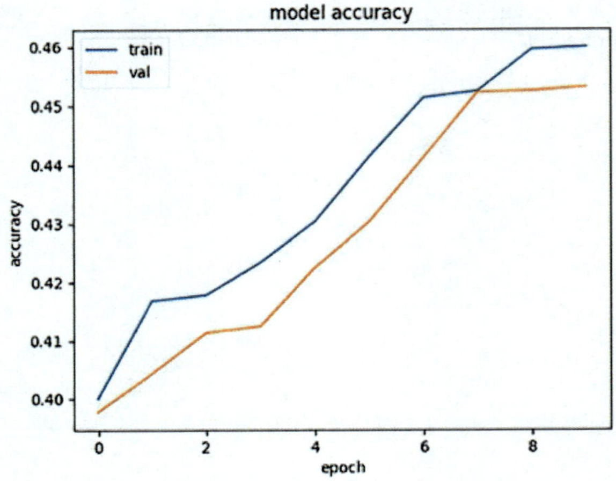


**Fig. 12** Accuracy per epoch for track 4

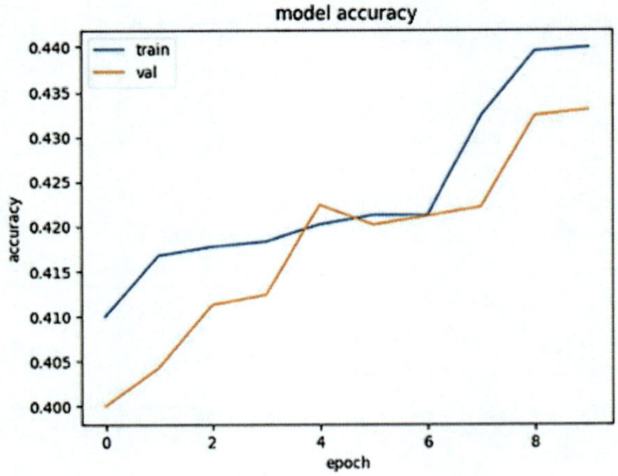


**Fig. 13** Accuracy per epoch for track 5

## 5 Conclusion and Future Works

This study proposed a concept of federated machine learning for making the car autonomous. Cars were trained in five different tracks to generate the individual model. Those individual models were combined with the help of server and that combined model was again updated through individual local training. We repeat this process for 10 times. We have used two different algorithm i.e. Federated Average and Scaffold Algorithm for model combining and analyzed the result. We have found that the overall accuracy decreases after using final combined model after 10 iteration. We found that this is because of client drift effect that is our combined model is biased towards certain tracks. After we use Scaffold algorithm for combining model, accuracy is almost similar in all tracks. We also found out that client drift effect from federated average method is reduced by using this algorithm.

The limitations of this study are that this research has not focused in increasing efficiency of the combined model. There is further research remaining on how to combine model efficiently to increase accuracy, since accuracy is very important factor for self-driving car. Further research is needed for developing new algorithm of model combination process that does not reduce the accuracy obtained from local training of clients.

**Acknowledgements** This research was supported by University Grants Commission (UGC), Nepal with collaborative research grant id: CRG-078/79-Engg-1, which was principally investigated by Dr. Babu R. Dawadi.

## References

1. Olaverri-Monreal C (2020) Promoting trust in self-driving vehicles. Nat Electron 3:292–294
2. Gu Z, Li Z, Di X, Shi R (2020) An LSTM-based autonomous driving model using a waymo open dataset. Appl Sci (Switzerland) 10(6):1–14. https://doi.org/10.3390/app10062046
3. Grigorescu S, Trasnea B, Cocias T, Macesanu G (2020) A survey of deep learning techniques for autonomous driving. J Field Rob 37(3):362–386. https://doi.org/10.1002/rob.21918
4. Pan X, You Y, Wang Z, Lu C (2017) Virtual to real reinforcement learning for autonomous driving. http://arxiv.org/abs/1704.03952
5. Chen J, Yuan B, Tomizuka M (2019) Model-free deep reinforcement learning for urban autonomous driving. http://arxiv.org/abs/1904.09503.
6. IEEE Computational Intelligence Society. International Neural Network Society, Institute of Electrical and Electronics Engineers, & IEEE World Congress on Computational Intelligence (2020 : Online). (2020). In: 2020 International joint conference on neural networks (IJCNN)
7. Mozaffari S, Al-Jarrah OY, Dianati M, Jennings P, Mouzakitis A (2022) Deep learning-based vehicle behavior prediction for autonomous driving applications: a review. IEEE Trans Intell Transp Syst 23(1):33–47. https://doi.org/10.1109/TITS.2020.3012034
8. McMahan HB, Moore E, Ramage D, et al (2016) Communication-efficient learning of deep networks from decentralized data. arXiv preprint arXiv:1602.05629
9. Reddi S, Charles Z, Zaheer, M et al (2020) Adaptive federated optimization. .arXiv preprint arXiv:2003.00295
10. Wang H, Yurochkin M, Sun, Y et al (2020) Federated learning with matched averaging. arXiv preprint arXiv:2002.06440
11. Tan M, Le QV (2019) Efficientnet: rethinking model scaling for convolutional neural networks. arXiv preprint. arXiv:1905.11946
12. He C, Annavaram M, Avestimehr S (2020) Fednas: Federated deep learning via neural architecture search. arXiv preprint. arXiv:2004.08546
13. Bernstein J, Wang Y-X, Azizzadenesheli K, et al (2018) SIGNSGD: compressed optimisation for non-convex problems. arXiv preprint. arXiv:1802.04434
14. Gupta O, Raskar R (2018) Distributed learning of deep neural network over multiple agents. J Netw Comput Appl 116:1–8
15. Vepakomma P, Gupta O, Swedish T, et al (2018) Split learning for health: Distributed deep learning without sharing raw patient data. arXiv preprint. arXiv:1812.00564
16. Lin T, Kong L, Stich SU, et al (2020) Ensemble distillation for robust model fusion in federated learning. arXiv preprint. arXiv:2006.07242

17. Hinton G,Vinyals O, Dean J (2015) Distilling the knowledge in a neural network. arXiv preprint. arXiv:1503.02531
18. Bucilua C,Caruana R, Niculescu-Mizil A (2006) Model compression. In: Proceedings of the 12th ACM SIGKDD international conference on Knowledge discovery and data mining, pp 535–541
19. Park J,Wang S, Elgabli A, et al (2015) Distilling on-device intelligence at the network edge. arXiv preprint. arXiv:1908.05895
20. He Y, Lin J, Liu Z, Wang H, Li LJ, Han S (2018) AMC: AutoML for model compression and acceleration on mobile devices. In: Ferrari V, Hebert M, Sminchisescu C, Weiss Y (eds) Computer Vision – ECCV 2018. ECCV 2018. LNCS, vol 11211, pp 784–800. Springer, Cham. https://doi.org/10.1007/978-3-030-01234-2_48
21. Zhang X, Zhou X, Lin M, et al (2018) SHUFFLENET: an extremely efficient convolutional neural network for mobile devices. In: Proceedings of the IEEE conference on computer vision and pattern recognition, pp 6848–6856
22. Iandola FN, Han S, Moskewicz MW, et al. (2016) Squeeze*N*et: Alexnet-level accuracy with 50x fewer parameters and<0.5 mb model size. arXiv preprint. arXiv:1602.07360
23. Wu B, Dai X, Zhang P, et al (2019) FBNET: Hardware-aware efficient ConvNet design via differentiable neural architecture search. In: Proceedings of the IEEE conference on computer vision and pattern recognition, pp 10734–10742
24. He C,Ye H, Shen L, et al (2020) Milenas: efficient neural architecture search via mixed-level reformulation. arXiv preprint. arXiv:2003.12238

# Predicting Foot and Mouth Disease in Thailand's Nakhon Ratchasima Province Through Machine Learning

**Wachirakan Sueabua and Pusadee Seresangtakul**

**Abstract** Every year, foot and mouth disease outbreaks occur in Thailand. Despite having a low fatality rate, the disease has a significant effect on numerous enterprises. Through the use of machine learning, the research herein forecasts foot and mouth disease in animals to alert farmers. In the study, support vector machines, decision trees, and artificial neural networks were utilized as prediction models. We studied and reviewed related works and collected nine risk factors for foot and mouth disease outbreaks. To analyze the appropriate models, data was collected from the Nakhon Ratchasima province from 2014 to 2020. The results showed very high accuracy; however, the precision, recall, and F1 were quite low. We determined that these results were due to imbalanced data sets. To improve the efficiency of the prediction models, we applied the synthetic minority oversampling technique (SMOTE) to solve the imbalance data set problem. The experiment results demonstrated that the decision trees model with nine factors together with adjusted imbalance data via the SMOTE technique outperformed the models constructed by ANNs and SVM with an accuracy of 98.86%.

**Keywords** Machine Learning · Neural Networks · Decision Trees · Support Vector Machines · Foot and Mouth Disease · Classification

## 1 Introduction

Thailand's Foot and Mouth disease (FMD) outbreaks have prompted the World Health Organization (WHO) to prohibit the exportation of diseased meat products to sterile countries; causing enormous economic losses [1–3]. FMD has been a common epidemic in Thailand for more than 50 years and usually occurs at the end of the

W. Sueabua · P. Seresangtakul (✉)
Department of Computer Science, College of Computing, Khon Kaen University, Khon, Kaen, Thailand
e-mail: pusadee@kku.ac.th

W. Sueabua
e-mail: wachirakan@kkumail.com



P. Meesad et al. (eds.), *Proceedings of the 19th International Conference on Computing and Information Technology (IC2IT 2023)*, Lecture Notes in Networks and Systems 679,
https://doi.org/10.1007/978-3-031-30474-3_5

rainy season through winter [4]. The Department of Livestock Development (DLD) provides free vaccinations for cattle and buffalo every year; however, the disease could not be contained, as many farmers did not accept the importance of vaccinations. Transport permits were eluded and sick animals were sold to markets. Risk factors for FMD outbreaks included the followings factors: animals were continually moving in and out of the farm; only partial vaccinations took place despite the entire herd being scheduled, disinfectants were not uniformly used, and animal purchases were made within a 10 km radius of a small farm [4] Machine learning is currently being applied to create forecasting models in various fields, such as product demand forecasting [5] and sales forecasting [6], as well as the likelihood of various diseases [7, 8]. Based on these conditions, appropriate factors were studied and modeled using machine learning for the classification of FMD. We, therefore, collected data on sick and hoofed animals from DLD, Nakhon Ratchasima, to make predictions through the artificial neural networks, support vector machines, and decision trees models. In the future, the improvement of FMD outbreak prediction will allow the Thailand's Department of Livestock better prepare and take preventive measures early for FMD outbreak.

## 2 Literature Review

Jia, Li, Tan and Xie presented the prognosis of HFMD in all provinces in China using machine learning. Data was collected from the department of hand foot and mouth disease in China. China center for disease control (CDC) and the Hong Kong department of health. People's searches for HFMD were also collected from the website. Using ridge regression, KNN, random forest, and recurrent neural networks models, the recurrent neural networks model is the best [9].

Nicholas, Denis and Labadin studied the relationship between climate change and hand, foot and mouth disease (HFMD) outbreaks in Sarawak. In this study, they collected the weather data, which consists of rainfall and mean surface air temperature between 2000 and 2008 in Sarawak. The evaluation results indicate that a relationship exists between the climate change and the HFMD outbreak [10].

Kheawsanam and Phonsawai presented that the incubation period for the virus is 2–14 days. Foot and mouth germs can survive for 14 days. Infected animals can infect other animals. Through secretions such as saliva, urine, feces, and semen, infected animals can infect other animals up to 4 days before they show symptoms. The disease is manifested by an infected animal by blisters in the oral cavity, mammary glands, and hoofs, the animal limp, salivation, gossip, milk decreases, and in young animals' death may occur. This study focuses on the study of risk factors in the outbreak of the disease [4].

Song and Lu presented a strong statistical tool for categorization, prediction, interpretation, and data manipulation is the decision trees approach. The key drawback is that, particularly with small data sets, it can be vulnerable to overfitting and underfitting. The generalizability and robustness of the resulting models may be constrained by this issue. Strong correlation between various potential input factors may lead to the selection of variables that enhance the model statistics but are not causally connected to the outcome of interest, which is another potential issue. As a result, care must be used when evaluating decision trees models and using their output to generate causal hypotheses [11].

Nyaguthii, Armson, Kitala, Sanz-Bernardo, Di Nardo, and Lyons presented Foot and Mouth disease (FMD) as a contagious disease in ruminants in Kenya, where epidemics occur year-round. From interviews with 220 small-hold farmers in Kenya, it was found that most farmers knew about FMD and were able to detect and identify sick animals. However, only 20% of the farmers vaccinated their animals. The risk factors causing FMD were the failure to bring animals for vaccination, free-range animal husbandry, the absence of measures to prevent outsiders from entering the farm, trading animals from animal markets, and the density of the animals in the farm [12].

The risk factors of FMD from these studies were used in our work. However, some factors were ignored such as free-range animal husbandry, the density of the animals in the farm, because of the limitation of our data source.

# 3 The Method

## *3.1 Research Method*

The research herein aims to construct a model to predict the breakout of FMD in animals in the Nakhon Ratchasima province. The research methodology is show in Fig. 1.

This study examines the yearly epidemic disease in cattle/buffalo (FMD) occurring in the Nakhon Ratchasima province. In this work, there are four steps including: data collection, data preprocessing, model construction, and model evaluation.

**Data Collection.** The outbreak risk factors of FMD have been previously determined; which included animal movement on the farm, non-vaccination of the entire herd, and the proximity (within a radius of 10 km) where animals were purchased [4]. The correlation between climate change and outbreaks of FMD in Sarawak, Malaysia had also been reported [10].

These factors were collected and used to construct the predictive model herein, as shown in Table 1. The variables included the year, month, district, rainfall, season, average temperature, animal movement, and the total number of animals within the farm and district, total amount of animals, the percentage of vaccinated animals, and the number of animals in the district. The data utilized herein were derived from Thailand's Department of Livestock and Meteorological Department, Nakhon Ratchasima.

Retrospective data were also collected over seven years from January through December of 2014; which outlined the animal husbandry sites of 32 districts, for a

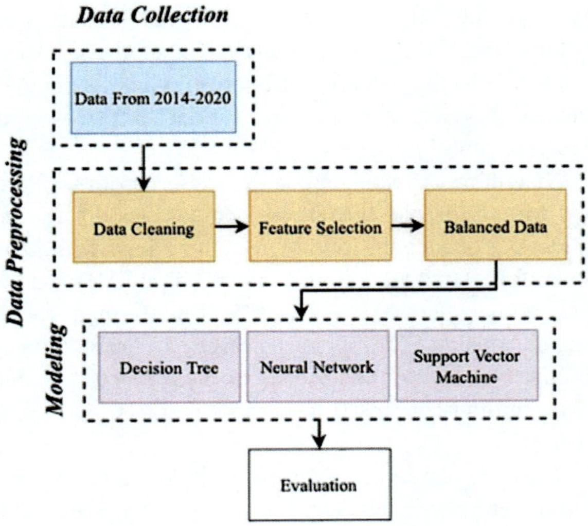


**Fig. 1** Research Methodology

**Table 1** Variables of foot and mouth disease (FMD) in the Nakhon Ratchasima data set

| Variable | Description | Example |
|---|---|---|
| Year | Year | 2014, 2015, 2016 |
| Month | Month | 1 = January<br>2 = February<br>3 = March etc |
| District | Animal husbandry site | 1 = Kaeng Sanam Nang<br>2 = Kham Talae So<br>3 = Kham Sakae Saeng |
| Rainfall | Rainfall (mm.) | 0, 201,467.9 |
| Season | Season | 1 = winter<br>2 = summer<br>3 = rainy<br>4 = summer rainy<br>5 = rainy winter<br>6 = winter summer |
| Temperature | Average temperature (per month) | 21.30 °C, 21.90 °C, 22 °C |
| Move_in | Numbers of transferred-in animals (each month) | 27, 29, 30 |
| Animal | Total number of animals (each month) | 3021, 3259, 3397, 108,108 |
| Vaccine | Percent of vaccinated animals (each month) | 10.30, 10.40, 12.70 |
| Market | Animal purchases in an animal market (each month) | 282, 320, 360 |
| Infect | Sick animal in the month | Yes/No |

total of 2,688 records. Through animal transport figures, the number of animals found infected with FMD in each district, for a total of 72 infected and 2,618 uninfected records, were recorded.

**Data Preprocessing.** The data processing techniques utilized herein consisted of three processes, which made the information suitable for processing with the machine learning model: data cleaning, data normalization, and feature selection.

***Data Cleaning.*** Some data utilized in this study were redundant and are referred to as duplicate data. For example, the precipitation data collected from the district of Phimai data set required the removal of all duplicate data. The average temperature and amount of rainfall for any given district may be incomplete, as the information is simply not collected. The district of Nakhon Ratchasima, examined herein, lacks a device to gauge rainfall, and was, therefore, estimated from the average rainfall of surrounding districts divided by the number of districts. The same approach was used for finding the average temperatures.

***Data Normalization.*** The purpose of data normalization is to change the numeric value in the dataset to a common scale, without distorting differences in the ranges of values or losing information [13]. In this study, data is normalized by applying Z-transform normalization, which is the distribution of data equal to 0 and a standard deviation equal to 1. A large difference in data values will result in greater loss [14], where data normalization was computed through the following equation.

$$z = \frac{x - \mu}{\sigma} \tag{1}$$

where z is the new value from the outcomes of normalization.
$x$ is the value before normalization.
$\mu$ is the mean
$\sigma$ is the standard deviation.

***Feature Selection.*** The factors associated with the outbreak of FMD consisted of the total number of animals, vaccination, animal husbandry, and temperature (monthly). Correlations were performed via the Weka program using the Best-First search method.

## *3.2 Model Construction*

Following the collection and manipulation of data, the next step was to bring the collected data into the forecast model through artificial neural networks (ANNs), the decision trees (DT), and the support vector machines (SVM). The ANNs were inspired by the brain's sensory processing model, which can be built by simulating a network of neurons simulated in a computer. Using an algorithm that learns the

**Table 2** Parameter settings for each model

| Model | Parameter setting |
|---|---|
| Neural Network | Training cycles = 1000<br>Learning rate = 1.0<br>Momentum = 0.1 |
| Decision Trees | Criterion = information gain<br>Maximal depth = 239<br>Confidence = 0.5<br>Minimal gain as NaN |
| SVM | Kernel type = Radial Basis Function<br>C = 100<br>Kernel gamma = 1.0<br>Kernel cache = 200 |

processes of neurons can enable learning networks to solve many types of problems [15]. The decision trees is a straightforward and simple-to-understand algorithm that can manage non-linear relationships between input and output variables, which makes it a strong contender for predicting FMD outbreaks [16]. The support vector machines is also a model of interest in forecasting research, as it is flexible in its adaptation and works well; particularly with complex data (multiple features) within a small sample size [17].

In this study, the incidence of FMD was predicted in animals using a classification method. Two groups, infected animals, and uninfected animals were examined following two steps: hyperparameter optimization and model validation. Hyperparameter optimization determines the best hyperparameters for the machine learning model employed, where every model has a number of adjustment options. The researchers adjusted the optimal parameter settings for each model via the RapidMiner program using grid search, as shown in Table 2. SVM in this experiment, C was adjusted using the following values: 100, 50, 10, 1.0, 0.1, 0.01 and the best result was 100.

In this study, we employed k-fold cross validation to evaluate the proposed system. Validating a model divides the data into k folds, where each fold must be randomly drawn to distribute data evenly for use in training and testing (research k = 10).

### 3.3 Model Evaluation

Evaluation is the assessment of a model according to its efficacy. In all three models, the confusion matrix was used to measure each model's accuracy. True Positive (TP) represents the predicted data that matches the actual data in the class under consideration, False Positive (FP) is the incorrectly predicted data for the class under consideration, False Negative (FN) denotes data that is incorrectly predicted as a class that is not considered, and True Negative (TN) is the predicted data that matches the true data in the class that is not considered [18], as show in Table 3.

**Table 3** The confusion matrix

| | Actually negative | Actually positive |
|---|---|---|
| Predicted Negative | True Negative (TN) | False Positive (FP) |
| Predicted Positive | False Negative (FN) | True Positive (TP) |

The confusion matrix is an important tool for evaluating prediction outcomes, in which accuracy, precision, recall, and F1 are used to assess the effectiveness of the predictions. Determined through Eqs. 2, 3, 4 and 5, *Accuracy* is the correctness of the predictions that match what actually happened, *Precision* denotes the correct (true) prediction, yet what is happening is not true. *Recall* is the accuracy of the prediction that will come true, compared to the number of times the events were predicted and true. Lastly, *F1* is the mean between precision and recall that measures the model's capabilities.

$$Accuracy = \frac{TP + TN}{TP + TN + FP + FN} \tag{2}$$

$$Precision = \frac{TP}{TP + FP} \tag{3}$$

$$Recall = \frac{TP}{TP + FN} \tag{4}$$

$$F1 = \frac{2(Precision * Recall)}{Precision + Recall} \tag{5}$$

## 4 Experiments and Results

In this study, we constructed models to predict outbreaks of FMD in the Nakhon Ratchasima province using three models: decision trees (DT), artificial neural networks (ANNs), and support vectors machine (SVM). In our experiments, we constructed the models using nine related factors: month, district, rainfall, season, temperature, move_in animal, vaccine, and market (as show in Table 1). There were also four selected factors: animal, vaccine, district and temperature. The results of each model's accuracy are presented in Tables 4 and 5.

The results indicated that the accuracies of all models were significantly high. However, infected prediction (the true positive value), precision, and recall were all quite low. This was a result of imbalanced data, in which data in one class is much higher than the amount of data in other classes. For example, in September 2020, Nakhon Ratchasima's Pak Chong district presented only a single case of FMD from a total of 108,108 animals. Typically, classification achieves the best performance when the model is trained using balanced datasets. If the data is not balanced, the

**Table 4** Experiment results of the related factors

| Model | TP (%) | TN (%) | FP (%) | FN (%) | Accuracy (%) | Precision (%) | Recall (%) | F1 (%) |
|---|---|---|---|---|---|---|---|---|
| ANNs | 53.67 | 99.42 | 0.58 | 45.90 | 98.22 | 53.67 | 99.42 | 61.68 |
| DT | 66.86 | 99.24 | 0.76 | 32.79 | 98.39 | 71.38 | 66.86 | 68.91 |
| SVM | 30.76 | 99.51 | 0.49 | 68.85 | 97.70 | 61.83 | 30.76 | 41.76 |

**Table 5** Experiment results of the selected factors

| Model | TP (%) | TN (%) | FP (%) | FN (%) | Accuracy (%) | Precision (%) | Recall (%) | F1 (%) |
|---|---|---|---|---|---|---|---|---|
| ANNs | 54.10 | 99.55 | 0.45 | 45.90 | 98.35 | 76.74 | 54.71 | 63.46 |
| DT | 63.93 | 99.02 | 0.98 | 36.07 | 98.09 | 66.46 | 64.00 | 63.93 |
| SVM | 39.34 | 93.27 | 6.73 | 60.66 | 91.83 | 13.71 | 39.14 | 20.34 |

classification will be biased towards the larger group [19]. Therefore, the data is balanced via the synthetic minority oversampling technique (SMOTE). The experiment results, outlined in Tables 6 and 7, showed that the decision trees (DT) produced the most significant prediction of FMD with the nine related factors. A correlation value was later determined through four selected factors: animal, vaccine, district and temperature. The results confirmed that the DT with optimized hyperparameters and SMOTE outperformed the other models (ANNs and SVM).

**Table 6** Experiment results of the nine related factors with SMOTE

| Model | TP (%) | TN (%) | FP (%) | FN (%) | Accuracy (%) | Precision (%) | Recall (%) | F1 (%) |
|---|---|---|---|---|---|---|---|---|
| ANNs | 98.97 | 97.73 | 2.27 | 1.03 | 98.35 | 97.75 | 98.97 | 98.36 |
| DT | 98.75 | 98.97 | 1.03 | 1.03 | 98.86 | 98.97 | 98.75 | 98.86 |
| SVM | 99.02 | 98.31 | 1.69 | 0.98 | 98.66 | 98.33 | 99.02 | 98.67 |

**Table 7** Experiment results of the four selected factors with SMOTE

| Model | TP (%) | TN (%) | FP (%) | FN (%) | Accuracy (%) | Precision (%) | Recall (%) | F1 (%) |
|---|---|---|---|---|---|---|---|---|
| ANNs | 95.90 | 90.06 | 9.94 | 4.10 | 92.98 | 90.68 | 95.90 | 93.18 |
| DT | 95.90 | 90.06 | 9.94 | 4.10 | 98.57 | 98.06 | 99.11 | 98.58 |
| SVM | 99.20 | 85.10 | 14.89 | 0.80 | 92.16 | 88.55 | 99.20 | 92.67 |

## 5 Conclusion

The objective of this study was to predict the incidence of FMD in animals located in Thailand's Nakhon Ratchasima province. Efforts were then made to increase farmer awareness and to help prevent the spread of FMD in Thailand by vaccinating animals on a regular basis. We concluded that the decision trees was most capable in handling imbalanced data via the SMOTE oversampling method containing nine variables: month, district, rainfall, season, temperature, move_in, animal, vaccine, and market. After finding a correlation value, four variables were then selected for further prediction study: animals, vaccine, district, and temperature. The decision trees was found to produce the most significant predictions using SMOTE oversampling with optimal hyperparameters with an accuracy of 98.86%.

**Acknowledgements** This paper was supported by College of Computing, Khon Kaen University, Khon Kaen, Thailand.

## References

1. Perry BD, Kalpravidh W, Coleman PG (1999) The economic impact of foot and mouth disease and its control in South-East Asia: a preliminary assessment with special reference to Thailand. OIE Revue Sci Tech 18:478–497 https://doi.org/10.20506/rst.18.2.1163.
2. Grubman MJ, Baxt B (2004) Foot-and-mouth disease. Clin Microbiol Rev 17:465–493. https://doi.org/10.1128/CMR.17.2.465-493.2004
3. Yano T, Premashthira S, Dejyong T, Tangtrongsup S, Salman MD (2018) The effectiveness of a foot and mouth disease outbreak control programme in Thailand 2008–2015: case studies and lessons learned. Vet Sci 5:1–13. https://doi.org/10.3390/vetsci5040101
4. Kheawsanam R, Phonsawai K Study of risk factor of foot and mouth disease outbreak in Nakhon Ratchasima Province. http://pvlo-nak.dld.go.th/research/ahg2.pdf. Accessed 3 Dec 2016
5. Moroff N, Kurt E, Kamphues J (2021) Machine learning and statistics: a study for assessing innovative demand forecasting models. Procedia Comput Sci. 180:40–49. https://doi.org/10.1016/j.procs.2021.01.127
6. Pavlyshenko BM (2019) Machine-learning models for sales time series forecasting. Data (Basel). 4:2–11. https://doi.org/10.3390/data4010015
7. Sharma G, Prabha C (2021) Applications of machine learning in cancer prediction and prognosis. In: Cancer prediction for industrial IoT 4.0: a machine learning perspective, pp 980–984. https://doi.org/10.1201/9781003185604-8
8 Wardeh M, Sharkey KJ, Baylis M (2020) Integration of shared-pathogen networks and machine learning reveals the key aspects of zoonoses and predicts mammalian reservoirs. Proc R Soc B Biol Sci 287:2–10. https://doi.org/10.1098/rspb.2019.2882
9 Jia W, Li X, Tan K, Xie G (2019) Predicting the outbreak of the hand-foot-mouth diseases in China using recurrent neural network. In: 2019 IEEE international conference on healthcare informatics (ICHI). pp 1–4. IEEE. https://doi.org/10.1109/ICHI.2019.8904736.
10 Nicholas M, Denis LA, Labadin J (2010) Preliminary investigation of the relationship between weather data and Hand-Foot-Mouth Disease cases in Sarawak. In: Proceedings - 2nd international conference on computational intelligence, modelling and simulation, CIMSim 2010, pp 373–378. https://doi.org/10.1109/CIMSiM.2010.53
11 Song YY, Lu Y (2015) Decision tree methods: applications for classification and prediction. Shanghai Arch Psychiatry 27:130–135. https://doi.org/10.11919/j.issn.1002-0829.215044

12 Nyaguthii DM, Armson B, Kitala PM, Sanz-Bernardo B, di Nardo A, Lyons NA (2019) Knowledge and risk factors for foot-and-mouth disease among small-scale dairy farmers in an endemic setting. Vet Res 50:1–12. https://doi.org/10.1186/s13567-019-0652-0
13 Martin, E (2021) Normalizing data. In: Martin E (ed) Experiments of the mind, pp 84–108. Princeton University Press. https://doi.org/10.2307/j.ctv1n1bs5c.11
14 Abdi H (2007) Z-scores. Encycl Measur Stat3:1055–1058
15 Cross SS, Harrison RF, Kennedy RL (1995) Introduction to neural networks. Lancet 346:1075–1079. https://doi.org/10.1016/S0140-6736(95)91746-2
16 Quinlan JR (1986) Induction of decision trees. Mach Learn 1:81–106. https://doi.org/10.1007/bf00116251
17 Steinwart I, Christmann A (2008) Support vector machines, pp 278–330. Springer, New York. https://doi.org/10.1007/978-0-387-77242-4
18 Shobha G, Rangaswamy S (2018) Chapter 8 - machine learning. In: Gudivada VN, Rao CR (eds) Computational analysis and understanding of natural languages: principles, methods and applications, pp 197–228. Elsevier. https://doi.org/10.1016/bs.host.2018.07.004
19 Chawla NV, Bowyer KW, Hall LO, Kegelmeyer WP (2002) SMOTE: synthetic minority over-sampling technique. J. Artif Intell Res **16:**321–357. https://doi.org/10.1613/jair.953

# Inspection of Injection Molding Process Improvement Using Simulation Techniques: A Case Study

**Patarida Loungklaypo and Srisawat Supsomboon**

**Abstract** Electric vehicles (EVs) are currently in high demand. The injection molding machine creates the connector, which is used to connect the electrical cables in the car. The injection molding inspection production volume is increasing as a result of the greater workload. The responsible department is dealing with a workload sharing issue since it must identify the specific machines for each worker. This study will identify, analyze, and improve the performance of injection molding inspections that happen as a result of quality control issues that cause worker utilization to be more than available time and unsuitable with the number of injection molding machines. The study's objective is to improve worker utilization and balance the injection molding inspection process's utilization without adding additional workers. The workload was analyzed using the line balancing theory. For line balancing with the transportation proportion model, Tecnomatix Plant Simulation was employed as a simulation and analysis tool. The simulation model's result will provide the number of shared machines for each worker to use in order to balance their workload.

**Keywords** Injection molding inspection · Worker utilization · Plant simulation · Line balancing

## 1 Introduction

The electric vehicle has become popular in Thailand in recent years; the manufacturing of electrical wire requires a certain amount of manual labor, and the connector has been widely used in the vehicle to connect the electronic system and cables. The big German automotive company produces the electrical connector injection molding to meet the requirements of high quality, reliability, and smaller size. As a result, workers are trained to identify any defects from the injection molding machine

P. Loungklaypo (✉) · S. Supsomboon
The Sirindhorn International Thai-German Graduate School of Engineering (TGGS), King Mongkut's University of Technology North, Bangkok 10800, Thailand
e-mail: 6209094860039@email.kmutnb.ac.th



P. Meesad et al. (eds.), *Proceedings of the 19th International Conference on Computing and Information Technology (IC2IT 2023)*, Lecture Notes in Networks and Systems 679,
https://doi.org/10.1007/978-3-031-30474-3_6

during the high volume of connector production that affect worker utilization. Moreover, to support the EV industry in the coming years, the company must increase the number of machines.

The company is dealing with higher utilization than available time due to an unmatched number of injection machines sharing, which affects the overall unbalanced workload. Therefore, there is a chance that a bottleneck will occur in the production line. As a consequence of the issue, the company's volume will increase in a few years because EV is popular in Thailand's manufacturing industry. As a result, not only must the quality of the injection molding process be improved, but the entire injection molding process as well. To that end, this study would assist the engineer in studying and evaluating the present and alternative options for improving injection molding process inspection.

The study aims to identify which is the bottleneck workstation and improve worker utilization, balance workload without adding any workers, and define the optimal number of machines to distribute the task. To improve the inspection process, the simulation is one method for detecting bottlenecks and studying production flow behavior. Additionally, the line balancing theory was applied to analyze the workload by the Siemens PLM software Tecnomatix Plant Simulation.

## 2 Background

Inspection is the process of inspecting sample parts from an injection molding machine for defects in appearance or dimension relative to standard instructions by taking some sample sizes before sending goods to the assembly line.

The injection molding inspection process flow chart is in the injection molding line to detect any defects in order to control quality issues from the machine immediately before the quality control department redetects them to ensure product specifications are met. Engineers assigned three workers to inspect the sample size, which was set so that they would receive Kanban from a running machine to detect the sample size, then walk to the machine to pick the sample and back to the inspection station. Following that, workers inspect the sample size in accordance with the engineers' specifications, then walk along to the machine to place the sample size and back to the inspection station to await an order from another Kanban. The process is shown in Fig. 1. At present, there is a station for three workers to inspect sample parts from 25 injection molding machines.

Figure 2 illustrates the worker distribution and machine layout with 25 machines. The processing time and throughput of each machine varies. In the model, the maximum throughput of each machine is 381,240 pieces.

This study will examine the current worker arrangement and attempt to assess workload usage at the bottleneck workstation as well as manage their workload through line balancing approaches.

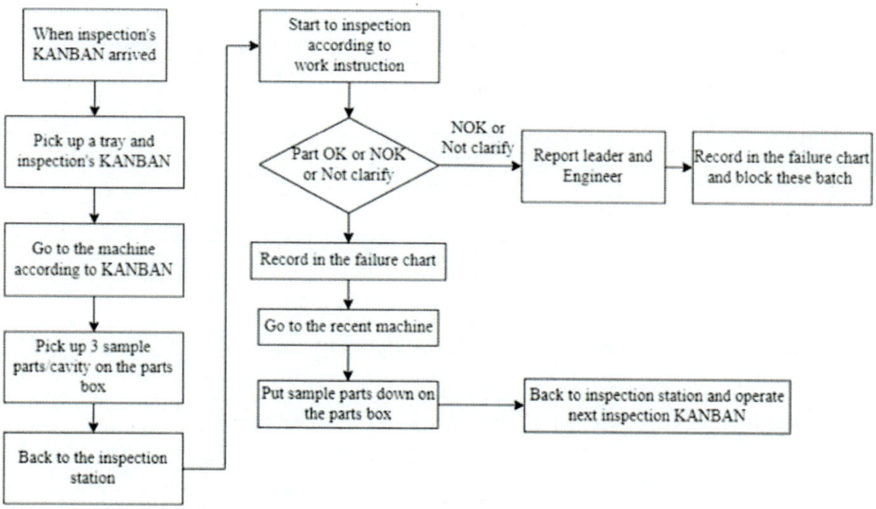


**Fig. 1.** Inspection of injection molding process

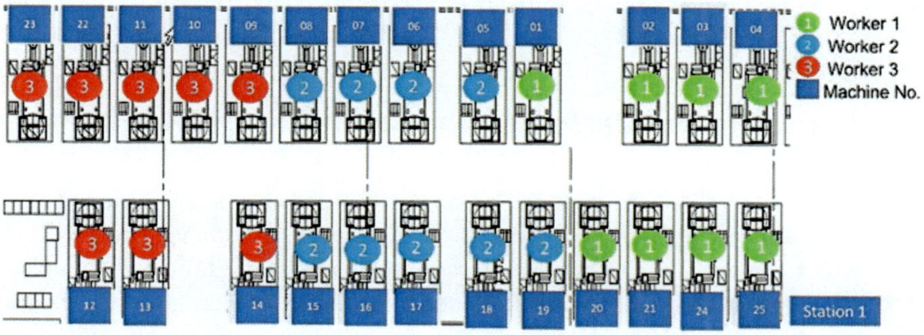


**Fig. 2.** Current machine layout and worker arrangements

## 3 Objective

The simulation is one tool for finding bottlenecks and evaluating production flow characteristics. The Siemens PLM Software Tecnomatix Plant Simulation also used the line balance theory to examine the workload. The results of the simulation model will identify the process bottleneck and remove unnecessary tasks, improve utilization, determine the appropriate number of machines employed by each worker, and establish a balanced task only for three people on twenty-five machines.

**Table 1** An example processing time of machines 1, 2 and 3 in minutes per day

| No | Machine1 | Machine2 | Machine3 |
|---|---|---|---|
| 1 | 117.31 | 171.81 | 95.34 |
| 2 | 120.71 | 169.60 | 95.48 |
| 3 | 121.04 | 168.38 | 95.40 |
| 4 | 121.04 | 168.74 | 97.12 |
| 5 | ... | ... | ... |
| 6 | ... | ... | ... |

## 4 Input Data Collection and Data Fitting

Data gathering is an important step in working with simulation tools. The model requires information from twenty-five stations, which include: processing times (inspection and recording time) and transition times during which workers carry the sampling size and return to the machine. During the data collection period, 30 samples of data are obtained using time-series analysis for each operation. Techniques- involving the use of the Minitab Input Analyzer Software evaluate all input to establish distribution types for each operation.

### *4.1 Processing Time Includes Inspection and Recording Time*

The primary challenge is finding skilled operators that can operate various types of injection equipment. Even if operators are discovered, calculating the new line balance problem with the present work force and those who are skilled is difficult.

The simulation modeling study described here is focused on how to create a reconfigurable simulation model in order to meet customer requirements while also improving system performance [1] Therefore, processing time is recorded by stop-watch for 30 samples of data per machine, including day and night shifts as shown in Table 1. A total of 25 machines have collected data on inspection, recording time, and other processes without carrying or walking time.

### *4.2 Transportation Time*

Walking speeds ranging from 0.06 to 1.3 m/s were identified using common assumptions [2]. Transportation time is a specified velocity, and the company's average walking speed standard is 0.9 m/s, multiplied by the distance between the inspection station and the injection machine (Fig. 4).

### 4.3 Input Distribution

30 samples of processing time data sets were evaluated by Minitab software in order to determine distribution types for every operation after being added to the time study data of 25 injection machines.[3] (see Table 2 and Fig. 3).

## 5 Simulation Model

Tecnomatix Plant Simulation, by Siemens PLM, was used to create the simulation model.[4] Using the automotive company's provided machine configuration (Fig. 1)

**Table 2** Statistic data generated by Minitab

| Machine | N | Mean | S.D | Min | Godness Test |
|---|---|---|---|---|---|
| 1 | 30 | 119.56 | 1.43 | 117.04 | Lognormal |
| 2 | 30 | 170.08 | 1.07 | 168.33 | Normal |
| 3 | 30 | 96.29 | 0.78 | 95.21 | Weibull |
| 4 | 30 | 119.05 | 1.26 | 117.12 | Normal |
| 5 | 30 | 157.01 | 1.36 | 154.02 | Normal |
| 6 | 30 | 136.12 | 1.03 | 134.20 | Lognormal |
| 7 | 30 | 96.84 | 1.86 | 93.17 | Normal |
| 8 | 30 | 78.33 | 1.59 | 76.59 | Gamma |
| 9 | 30 | 230.02 | 2.02 | 226.01 | Lognormal |
| 10 | 30 | 107.52 | 1.51 | 105.05 | Normal |
| 11 | 30 | 285.13 | 1.64 | 282.25 | Lognormal |
| 12 | 30 | 162.21 | 1.52 | 159.24 | Normal |
| 13 | 30 | 187.07 | 1.94 | 184.33 | Normal |
| 14 | 30 | 143.85 | 2.33 | 140.04 | Lognormal |
| 15 | 30 | 148.80 | 1.55 | 146.09 | Weibull |
| 16 | 30 | 88.83 | 1.68 | 85.38 | Normal |
| 17 | 30 | 141.93 | 1.78 | 138.03 | Gamma |
| 18 | 30 | 215.14 | 1.22 | 213.24 | Weibull |
| 19 | 30 | 68.20 | 1.75 | 65.11 | Normal |
| 20 | 30 | 193.65 | 1.51 | 190.40 | Gamma |
| 21 | 30 | 300.70 | 1.84 | 297.08 | Normal |
| 22 | 30 | 57.10 | 1.52 | 54.28 | Gamma |
| 23 | 30 | 66.89 | 1.80 | 63.58 | Lognormal |
| 24 | 30 | 40.88 | 1.96 | 36.40 | Normal |
| 25 | 30 | 45.12 | 2.29 | 40.16 | Weibull |

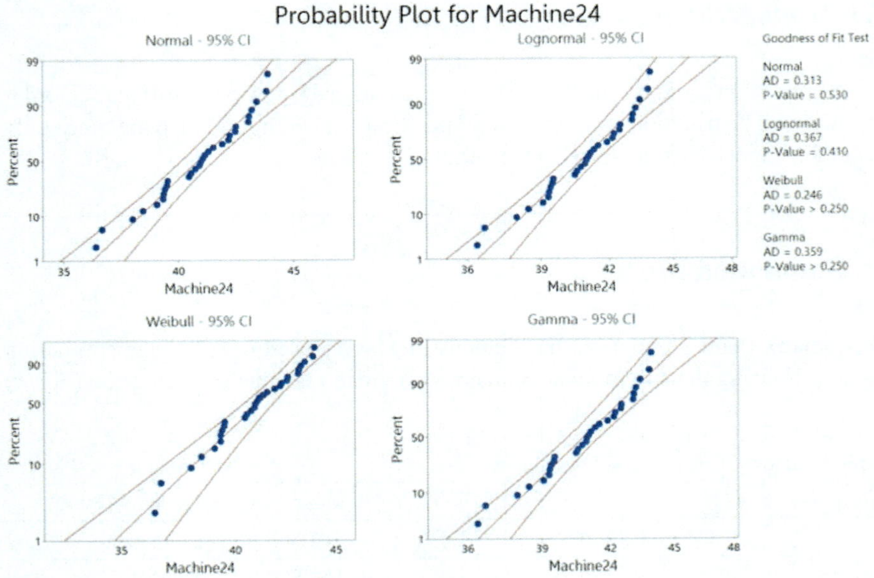


**Fig. 3** Machine 24: an example of determining the most appropriate distribution fit

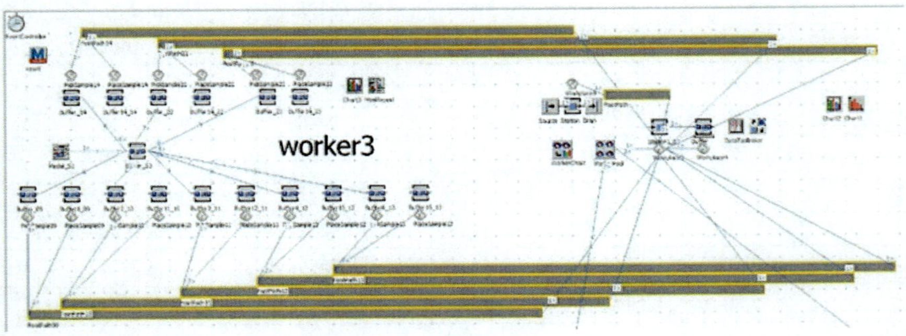


**Fig. 4** Woker3: an example of current process modeling

and three workers arrangements, where each worker was assigned a different number of machines, which are twenty-five, and the distance between each machine and an inspection machine varied, we can estimate how long the model's production will take. The company spends 720 min each day working, including breaks, and this research represents discrete-event modeling. Each order begins with a blank piece of the production line. The problem is defined by the following assumptions:

As a consequence, there are no interruptions in manufacturing. Time required for setting up is not included in. Since in a real system, this initialization phase is often finished by the end of the working day (645 min including breaks). All process times for injection machines include negligible breakdowns, even if no maintenance is conducted throughout the working period. The maximum throughout is applied to each machine to seek maximum worker utilization in the actual system.[5]

## 6 Bottleneck Identification

In terms of the current process, worker 1 has the shortest distance, followed by workers 2 and 3, as shown in Fig. 7. As a result of the long distance for walking and the unbalanced number of machines assigned to workers, the bottleneck position occurs at the peak workload at worker 2 and 3. The output data can help with scheduling and designing the new models, which will be shown in the following topic.

## 7 Experimental Design

The existing inspection method with three workers has a high utilization rate and an unbalanced workload, according to the workers. Because of this, the line balancing method is employed to consider alternative solutions for achieving the desired outcome without affecting product quality. Next, two models focused on developing new models were held. Improvements in utilization may have been achieved through the employing of different models, such as those considered in this study, which included rearranging the machine's layout, assigning specific tasks to each worker, adding a new inspection station, and eliminating the bottleneck.[6] As a result, by creating a new station, these two models are intended to shorten transportation time. Model A depicts a new station with the current machine configuration. Additionally, as demonstrated in Fig. 8, model B improves on model A in that the workload is not balanced. As a consequence, Model B will reallocate the machine layout with the workers by sharing the task of worker 2 across workers 1 and 3.

### *7.1 Model A: Current Machine Layout and 2 Inspection Stations*

Model A (shown in Fig. 5) is a representation of the current production process that adds an inspection station to improve the workload's utilization. As a result, transportation time between the inspection machines and the two inspection stations is cut short only for workers 2 and 3.

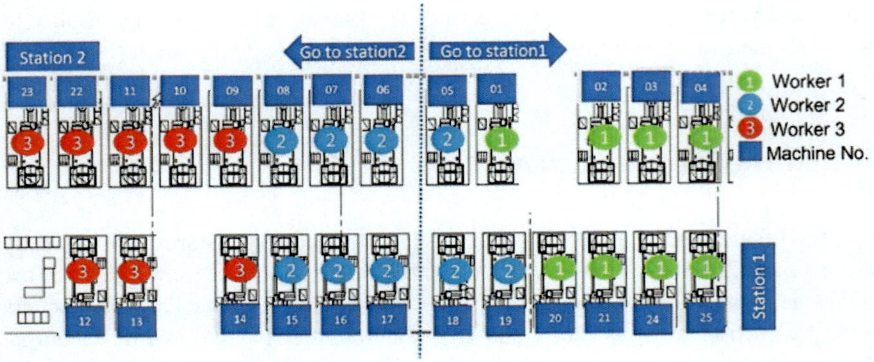


**Fig. 5** Current machine layout with adding a new inspection station

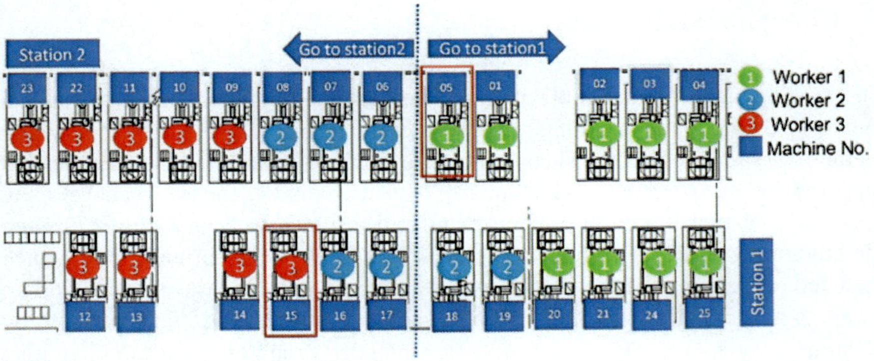


**Fig. 6** Rearranged machine layout with adding a new inspection station

## *7.2 Model B: Rearranged Machine Layout and 2 Inspection Stations*

Model B (shown in Fig. 6) has two inspection stations, and Worker 2's workload is divided between Workers 1 and 3 so that Worker 1 inspects machine 5 and Worker 3 inspects machine 15. The bottleneck status of worker 2 has improved because of this modification in layout machine allocation and reduction in transportation time.

# 8 Model Validation

To achieve model reliability, statistical validity requires a key factor between the real system and output performance. If there is no statistically significant difference

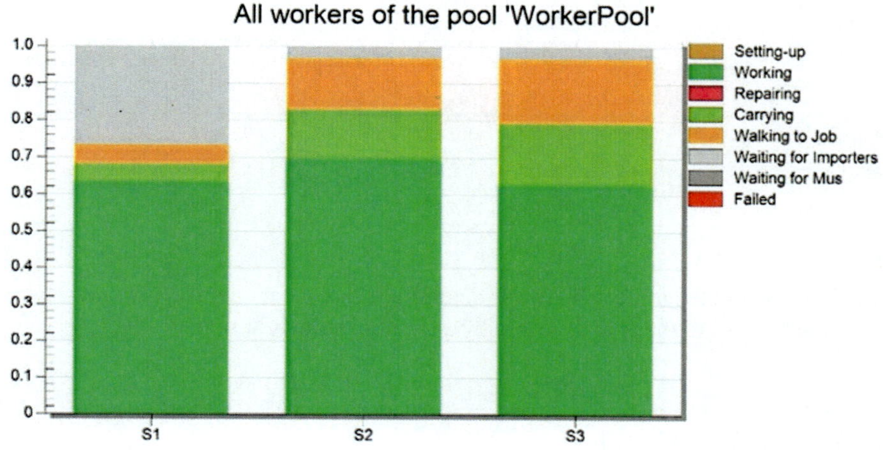


**Fig. 7** Utilization of three workers with current process

**Fig. 8** Utilization of three workers for each model

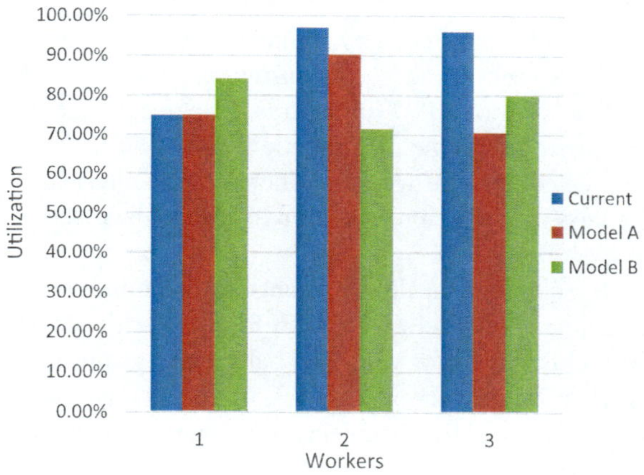


between the data sets, the model is considered acceptable. The modeling is rejected if there is a statistically significant difference in the opposite direction [1].

As a consequence, the statistical data were used to the simulation software, Tecnomatix Plant Simulation, to model the system and to demonstrate the current inspection process. The results were then generated with worker's utilization.

The t-test method is employed on sets of data separately at a level of significance, with a value of alpha of 0.05 being commonly applied. The modeling data are statistically similar using hypothesis testing, which involves the establishment of a null hypothesis, which implies that the data sets are statistically related, and the alternative hypothesis setting, which implies that the data sets are not statistically related. The significance level was determined using a t-test. The resultant t value of 0.332 was between the significant values of −2.571 and +2.571, with 5 degrees of freedom

**Table 3** The model results of three workers

| | Utilization | | | Average of utilization | Line balancing efficiency |
|---|---|---|---|---|---|
| | Worker1 | Worker2 | Worker3 | | |
| Current | 74.87% | 97.04% | 96.10% | 89.33% | 92.06% |
| Model A | 74.84% | 90.16% | 70.46% | 78.49% | 87.06% |
| Model B | 84.13% | 71.40% | 79.95% | 78.49% | 93.30% |

and a 0.05 level of significance. It was determined that the model was accurate since it did not vary from the actual system by a significant margin statistically.

## 9 Results and Conclusion

After the validation process was done, different decision options were evaluated for the design or reconfiguration of the inspection process. As shown in Model B, the bottleneck is improved by adding a new inspection station and worker arrangement.

As is seen in Fig. 7, the current process has high utilization for workers 2 and 3. That's almost 100% utilization. Due to customer specifications in quality control, the processing time of inspection cannot be reduced because standards are set for every skill level of workers. Therefore, an alternative model to improve the current process is adding a new station to reduce the distance between the machine and the inspection station.

Even though the development of utilization is reduced by adding a new inspection machine, as Model A shows in Fig. 8, workloads are still unbalanced. As a result, Model B is held for workload balancing by splitting the assigned machine between worker 2 and workers 1 and 3. In Table 3, a new station can improve 10.84% of utilization, and line balancing efficiency can improve by 6% by dividing worker 2's workload.

However, utilization may be improved by cutting down on the amount of time spent processing, inspecting, or recording information. This will make it feasible for more skilled workers to be recognized and trained. In addition to the workload, greater balance can be obtained if the worker arrangement is generated using the line balancing approach to identify the suitable number of machines to use with three workers. This could be done to determine the optimal number of machines. Additionally, various alternative worker arrangements can be implemented by plant simulation software to adjust any state in a real system, and engineers can avoid any unexpected situations by evaluating the results of the simulation model.

## References

1. Güner GM, Ünal C (2008) Line balancing in the apparel industry using simulation techniques. Fibres Text East Europe 16(2):75–78
2. Kay F, Marcus AB, Shawn T (2006) Another look at pedestrian walking speed. Transp Res Rec 1982(1):21–29
3. Minitab 21 Support. https://support.minitab.com. Accessed 25 Nov 2022
4. MES, Martijn. Simulation modelling using practical examples: a plant simulation tutorial. Mes M, 2017.
5. Chung CA (2003) Simulation modeling handbook: a practical approach. CRC Press
6. Kikolski M (2016) Identification of production bottlenecks with the use of plant simulation software. Eng Manag Prod Serv 8(4):103–112

# An End-to-end Framework to Harness Knowledge Graphs for Building Better Models from Data

**Sasin Janpuangtong**

**Abstract** Readily available data and software tools have turned "data analytics" into a game anyone can play. But genuine, serious modeling and analysis demands prudence; novice model builders may have inadequate knowledge (of either the domain or methods) to make judicious decisions particularly on which input features should be used to learn good models. Consequently, the practice, so called "garbage in, garbage out", would be entertained. This work introduces an end-to-end model building framework that makes use of background and semantic knowledge in the form of a knowledge graph to automatically determine the relevance of features from certain structural information, helping produce models with semantically pertinent features rather than mere correlations or (potentially spurious) patterns within the data. This proposed framework enables model builders (even non-experts) to build models from data informed by knowledge latent within any knowledge graph. We detail techniques used in our implementation by making use of an existing knowledge graph. To validate our approach, two models were built from a given dataset using our system, and then prediction errors of these models were examined. The results show that the proposed framework helps in improving model generalization, leading to a better prediction performance of learned models on unseen examples.

**Keywords** knowledge graphs · semantic feature selection · machine learning process

## 1 Introduction

The availability of data on the web, along with easy-to-use commodity software for regression and classification, makes it increasingly straightforward for non-expert users to analyze substantial volumes of data. Consequently, so called data dredging [1]—finding correlations amongst a large set of variables with scant regard for

S. Janpuangtong (✉)
Department of Science and Technology, Songkhla Rajabhat University, Songkhla, Thailand
e-mail: sasin.ja@skru.ac.th

 
P. Meesad et al. (eds.), *Proceedings of the 19th International Conference on Computing and Information Technology (IC2IT 2023)*, Lecture Notes in Networks and Systems 679,
https://doi.org/10.1007/978-3-031-30474-3_7

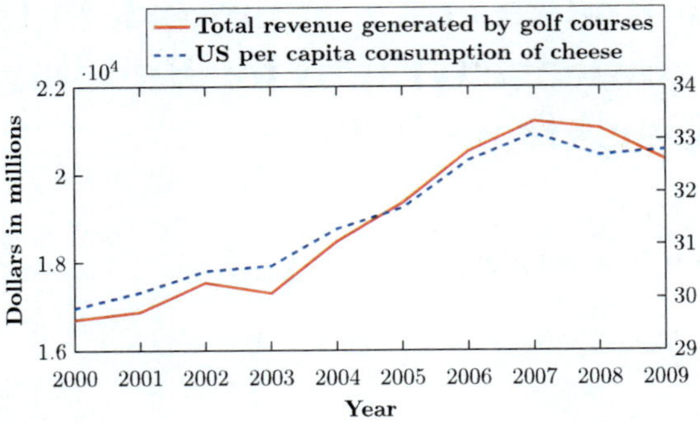


**Fig. 1** The dangers of data dredging: an example of spurious correlations within data. The plot shows per capita consumption of cheese in the US along with total revenue generated by golf courses in the US. The two are highly correlated ($\rho = 0.989$, source: http://www.tylervigen.com).

their meaning—has been democratized and anyone may use seemingly sophisticated tools to arrive at fallacious conclusions through indiscriminate use of data. As a motivating example, consider Fig. 1 which shows that data for consumption of cheese per capita in the US is highly correlated with total revenue generated by golf courses. Logically, we know that greater consumption of cheese does not cause people to head to the links for a whip around the green (nor vice-versa). But without the capacity to distinguish spurious correlations from others, learning methods are unfortunately prone to pick up this correlation [2].

Online data not only provide a vast set of potential features with which to build interesting models but also introduce many redundant and irrelevant features, which impose challenges to learning methods in separating the signal from the noise. Especially whenever the number of input features is large with respect to the number of training examples, chance correlations and spurious patterns can lead to specious models, which may be overfit and difficult to interpret [3]. Thus the onus ultimately falls on the person building a model to be judicious: they must exercise caution to avoid the hazards of data dredging. That is, we argue that without taking semantic and background knowledge into account, the traditional learning process inevitably and unconsciously leads model builders to exercise the *garbage-in, garbage-out* practice. Yet, current software tools do little to shield them from such potential pitfalls.

One broad approach used to allay this fundamental problem is for the model builder to use their knowledge of the domain in order to exercise discretion in assessing the relevance of each input feature. This approach is shown in Fig. 2a: the model builder interprets features not merely as vectors of data, but associates them with *concepts* within a mental framework, providing meaning for the feature. The model builder then performs a cognitive process using semantic knowledge to determine the redundancy and relevance of features. This avenue is, of course, only open to those who have suitable expertise and is, generally, unavailable to anyone. Moreover, even for domain experts, manually selecting relevant features from a large dataset can be time consuming.

The contributions of this work are three folds: (1) A new model building process that incorporates background knowledge is introduced in order to support non-expert model builders in building high prediction performance models. (2) Rather than relying on background knowledge from the person building the model, an end-to-

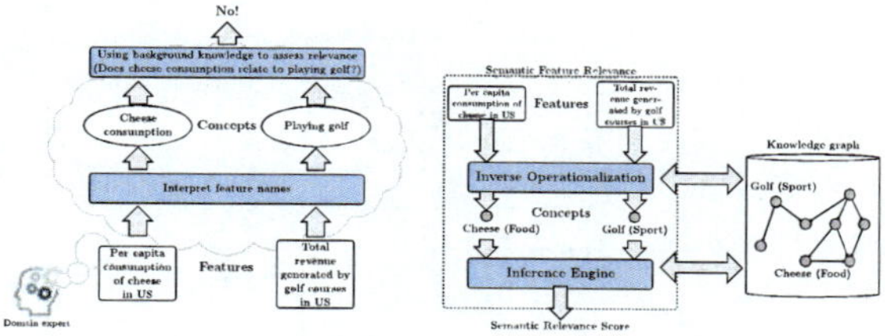


(a) (b)

**Fig. 2** Comparing the proposed feature selection framework to what experts do : (a) Handling of the previous scenario: the domain expert shrewdly rejects models based on these features due to their semantics. We automate an imitation of this cognitive process. (b) The proposed feature selection approach harnesses structural information in a given knowledge graph to assess the relevance between two features. Given feature names, the inverse operationalization component searches for a corresponding concept in the knowledge graph to represent each feature. Next, the inference engine quantifies the relevance score between the concepts using graph's structural information.

end model building framework that exploits high-level knowledge in the form of a knowledge graph to automatically estimate the relevance of features from a certain structural information within the graph is proposed. (3) An initial system is developed from the proposed framework by exploiting existing knowledge graphs and data available on online data sources to evaluate the underlying ideas.

## 2 Semantic Feature Selection

The main component of the proposed framework is semantic feature selection, illustrated in Fig. 2b, which makes use of structural information in a given knowledge graph to automatically assess the mutual relevance of two features. By imitating the cognitive process performed by a domain expert shown in Fig. 2a, this performs in two steps. The first step, *Inverse Operationalization*[1], operates under the idea that a feature is a measurement of an abstract concept. This step uses the given feature name to look for a corresponding concept within the knowledge graph. The concept—feature-name association is used to express the semantics of the feature. For simplicity, we assume that a feature is associated with one concept.

The second step, performed by the *Inference Engine*, takes two concepts obtained from the first step and computes a numerical score reflecting the mutual relevance of the pair using structural information within the knowledge graph. Relevance is computing using a graph distance between concepts, in which concepts are less likely to be relevant if their are further apart in the knowledge graph. Obviously, performance of this step depends directly on the accuracy of the outputs from the former step.

[1] Operationalization refers to a process that scientists usually perform to transform a concept to a corresponding measurement. For example, the concept of "Obesity" can be measured by the body mass index.

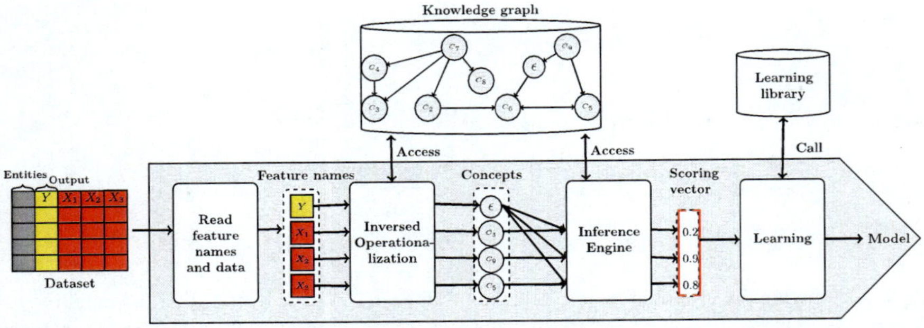


**Fig. 3** The learning framework. Given a dataset and a knowledge graph, it retrieves a concept to associate with each input feature/attribute. A semantic relevance metric between a concept representing the output column ($Y$) and a concept representing each attribute column is determined and then used to select features for building a model via a selected learning method.

## 3 Model Building Framework

Combining the preceding feature selection metric with existing learning algorithms, we propose an end-to-end learning framework (see Fig. 3) which automatically assesses relevance of each feature in a given dataset, and then chooses only relevant features using a given threshold while learning a model. Using this framework to build a model, a tabular input dataset is assumed to be structured in the following format: (1) The first column contains a list of entities over which a model is applied. We call this the entity column. (2) Other columns, called attribute columns, contain numeric data from attributes of entities. (3) One of these attribute columns is used as a quantity that the model aims to predict. This is called the output column. (4) Each attribute column contains meaningful column name at the first row.

The following is an example use case scenario. An inexperienced model builder, obtains data about countries in Excel format from CIA world factbook website[2]). Suppose the first column of this dataset contains a list of countries and other columns contain numeric data of various attributes of a country with a meaningful column name associated with them. From this dataset, the model builder aims to build a model for predicting GDP real growth rate of countries to understand the influence of other factors (various attributes of countries) on a country's monetary wealth, but without detailed background knowledge of each of the attributes in the input.

In our framework, the model builder inputs the dataset and interacts with the web interface to build the desired model. The framework performs inverse operationalization to associate each column with a corresponding concept in whichever knowledge graph the user selects. Let $\epsilon$ denote the concept obtained which represents the output column, and a concept representing attribute column $i$ be $C_i$ for every $i$. Next, the inference engine component determines the relevance between $\epsilon$ and $C_i$. The relevance score of $C_i$, calculated in this step, is assigned to the corresponding column.

[2] https://www.cia.gov/library/publications/download/.

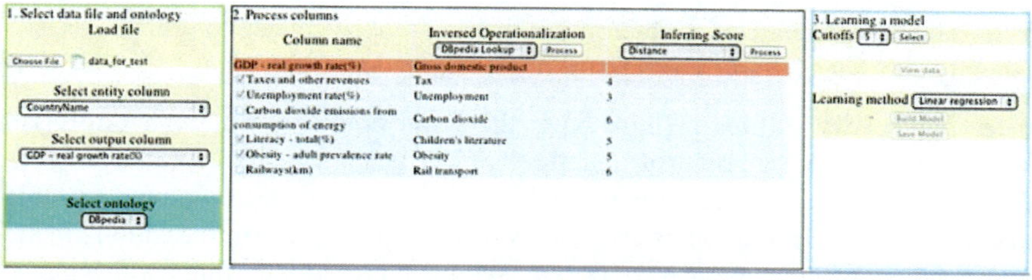


**Fig. 4** User interface of the system's prototype: On the left hand side, the user selects an input dataset, specifies the entity and output columns, and also chooses a knowledge graph. Calculating relevant score for each feature is performed on the middle part. The user then specifies the number of features used for learning and chooses a learning method to build a model.

Attribute columns are then ordered regarding of their score. The user selects a learning method and also specifies the number of features used to construct a dataset for learning, then a model is built from the dataset using the chosen learning method.

# 4 DBpedia Implementation

We implemented a specific system using DBpedia [4], the knowledge graph counterpart of Wikipedia. We chose this one because it is easy to access and provides general knowledge for a variety of problem domains, which allows us to test our system on different use cases. The user interface of the system implemented in this section is shown in Fig. 4.

## *4.1 Inverse Operationalization*

Inverse operationalization is an entity recognition task, in which input terms are annotated by pre-defined concepts or categories. Existing entity recognition techniques can be used to implement this component. Here, we propose two simple implementations working specifically with DBpedia: using its Lookup Service[3] and using Wikipedia tables project [5].

**Using DBpedia Lookup Service.** DBpedia provides this service for looking up concepts from input keywords. DBpedia concepts whose labels or referring anchor texts used in their associated Wikipedia articles match the keywords are retrieved. The resulting concepts are ranked based on the number of in-links pointing from other Wikipedia articles to the concepts' associated articles. Implementing inverse operationalization component using this service can be done by firstly processing a column name and then call the service with the processed column name as a query

[3] http://lookup.dbpedia.org/api/search/KeywordSearch.

string. The component could either return the first ranked concept or allow a user to manually choose a suitable concept for improving its accuracy.

**Using Wikipedia Tables.** Wikipedia gathers information from many different sources, a significant amount of quantitative data is specifically captured in tabular form. The key idea behind using wikipedia tables is that a column in a table published as part of a Wikipedia article may be found to be similar to a column in the input dataset. Given a column $c$, searching Wikipedia for articles that publish tables containing a column whose column name is similar to the column name of $c$ allows the system to match $c$ with corresponding topics.

In order to implement the inverse operationalization component from this project, two additional inverted indexes were constructed. The first index, denoted $I_t$, indexes tables by terms appearing in their titles and column names. Occurrences of term $t$ are counted to use as the term frequency value ($tf$) corresponding to a table. The inverse document frequency value of $t$ is calculated by $idf_t = \log N/df_t$, where $N$ is the total number of Wikipedia tables and $df_t$ is the number of tables in Wikipedia that contain term $t$. The second index, denoted $I_e$, indexes tables by terms appearing in the entity column of the input dataset, which we denote each of these terms by $e$.

Suppose the input dataset contains a set of terms $E$, in the entity column and each of its attribute column, $c$, contains a set of terms, $T_c$, as its column name. For each $c$, the system looks for Wikipedia tables, which have their titles or column headers textually matched with $T_c$ and also contain a column whose data matches (textually) $E$. To this end, the indexes $I_t$ and $I_e$ are used to calculate a similarity score for each Wikipedia table, $\tau$, with respect to $T_c$ and $E$. This score is calculated via the following formula:

$$\text{table\_similarity}_\tau = \alpha \left( \frac{\mathbf{V_c} \cdot \mathbf{V}_{\tau_c}}{|\mathbf{V_c}||\mathbf{V}_{\tau_c}|} \right) + (1 - \alpha) \left( \frac{\mathbf{V_e} \cdot \mathbf{V}_{\tau_e}}{|\mathbf{V_e}||\mathbf{V}_{\tau_e}|} \right). \tag{1}$$

where $\mathbf{V_c}$ and $\mathbf{V}_{\tau_c}$ are vectors, with one component representing each indexed term in $I_t$ and containing tf-idf weighting calculated from the index. The first vector is constructed directly from $T_c$, while the latter comes from terms appearing in $\tau$ (i.e., its title and column headers). For the second component, $\mathbf{V_e}$ and $\mathbf{V}_{\tau_e}$ are vectors, with one component representing each indexed term in $I_e$ and containing tf-idf weighting calculated from the index. The first vector is constructed directly from terms in $E$, while the latter comes from entity terms appearing in $\tau$. We use $\alpha = 0.7$, placing moderately more weight on the first component. The tables are sorted on their score and the corresponding article for each table is queried from the database. Next, the corresponding concept for each article is acquired from DBpedia. One concept is selected to represent $c$ either, heuristically, by using the highest ranked concept or by allowing a user to select a suitable one from among the top-ranking concepts.

## 4.2 Inference Engine

Given a concepts $\epsilon$ and $c_i$ representing the output column and an attribute column $i$ in the dataset, the inference engine calculates a relevance score between $\epsilon$ and $c_i$. It performs this calculation by using the taxonomic hierarchy within the knowledge graph, since closely related concepts are organized into the same category (category-concept relationships) and closely related specific categories are organized into the same broader category (category-category relationships). Let's represent the taxonomic hierarchy by a graph, $G = (V, E)$, where $V$ is a set of categories and concepts and $E$ is a set of edges encoding two types of relationships in the hierarchy through different predicates: (i) http://purl.org/dc/terms/subject encoded category-concept relationships, and (ii) http://www.w3.org/2004/02/skos/core#broader encoded category-category relationships. Using $G$, we proposed two approaches to determine the score. Both approaches start by retrieving corresponding categories of $\epsilon$ and $c_i$ using the first type of relationships. If we denote these retrieved categories by $C_\epsilon$ and $C_i$, then the category-category relationships are used differently by each approach.

**Edge Counting.** This approach was originally proposed to quantify similarity between two strings by counting edges between concepts representing those strings in a taxonomic graph [6, 7]. For each category $\delta_\epsilon \in C_\epsilon$, the shortest path to each category $\delta_i \in C_i$ is calculated. The minimum shortest path between categories in $C_\epsilon$ and $C_i$ is used to quantify the relevance between $c_\epsilon$ and $c_i$,

$$\text{score}_i = \min_{\delta_\epsilon \in C_\epsilon, \delta_i \in C_i} \textbf{shortest-path}(\delta_\epsilon, \delta_i). \tag{2}$$

**Information Content.** The main drawback of the edge counting approach is the assumption that edges in the taxonomic graph represent uniform distances. In real taxonomies edges can be heterogeneous and some sub-taxonomies are much denser than others. To address this problem, Resnik [8] proposed an information-based approach to measure the similarity of two concepts in a taxonomy. Applying his approach, each category, $\delta$, in $G$ is associated with $p(\delta)$, which is the probability of encountering subcategories of $\delta$. Such that the information content of $\delta$ can be calculated by $\mathit{Inf}(\delta) = -\log p(\delta)$, where

$$p(\delta) = |\text{nodes of the sub-tree rooted with } \delta| / |\text{nodes in } G|.$$

As the probability of $\delta$ increases, its informativeness decreases. The more abstract the category $\delta$, the lower its information content; a broad category conveys less information than a specific one. Intuitively, the more information two concepts share, the more similar they are and so the information shared by two concepts indicative of the information content of the most specific category that subsumes both $c_o$ and $c_c$, such that

$$\text{score}_c = \begin{cases} \max\limits_{\delta \in C_o \cap C_c} \mathit{Inf}(\delta) & \text{if } C_o \cap C_c \neq \emptyset, \\ \max\limits_{\delta_o \in C_o, \delta_c \in C_c} \max\limits_{\delta \in S(\delta_o, \delta_c)} \mathit{Inf}(\delta) & \text{otherwise.} \end{cases} \tag{3}$$

The second case is used when there is no category that directly subsumes both $c_o$ and $c_c$. Set $S(\delta_o, \delta_c)$ is a set of categories that subsumes both $\delta_o$ and $\delta_c$.

## 5 Evaluation

Our evaluation sets out to answer the following questions: (i) Does using the knowledge graphs help improve generalization of the models produced, and are the features selected by our framework better than features selected using pure correlations in data? (ii) How do the various techniques used in implementing components of the framework impact the quality of the models that result? The implementation described in the previous section was evaluated on different problem domains using a given dataset in order to answer these questions.

The given dataset contains a variety of attributes associated with countries collected from the CIA world factbook, in which countries are treated as the entities. Before using this dataset, the issue of missing data was addressed by removing columns and rows where 70% of their data were missing. Then, we manually filled in the remaining missing data for each column with the mean value of data in that column. After completing this data cleaning processes, the dataset has 179 rows and 140 columns.

Various models were built from this dataset as we implemented four different versions of the system using different combinations of the proposed methods for inverse operationalization and inference engine components as follows: (1) **LookUpEdc:** DBpedia lookup and edge counting, (2) **LookUpInf:** DBpedia lookup and information content, (3) **TablesEdc:** Wikipedia tables and edge counting, and (4) **TablesInf:** Wikipedia tables and information content. Besides comparing models produced from the different versions of the system, we also implemented a baseline system, called **Univariate**, to compare with all four systems implemented from our framework. This baseline one simply applies univariate feature selection technique to select a given number of features by using only correlations measured in the dataset. Using different implementations of the inverse operationalization component yielded different sets of concepts with differing precision for each feature. We obtained an average precision of 0.42 when using the DBpedia lookup service, while using Wikipedia tables not only gave us higher average precision at 0.61 but the concepts returned for each feature were all specific to countries.

Models of two different problems: GDP and poverty were built from this dataset. A column corresponding to the attribute we wish to predict was selected to use as the output column. These two problems were selected partly because they yielded high precision concepts from the inverse operationalization step. In order to examine the phenomenon of overfitting, we constructed two sets of cutoffs corresponding to two versions of the inference engine, where we use $\Gamma_e$ for edge counting and $\Gamma_i$ for information content. These cutoffs help specify a number of features used to build a model. Given $\gamma_e \in \Gamma_e$, a column is selected if its edge counting score $\leq \gamma_e$; whereas given $\gamma_i \in \Gamma_i$, a column is selected if its information content score $\geq \gamma_i$. Hence, the

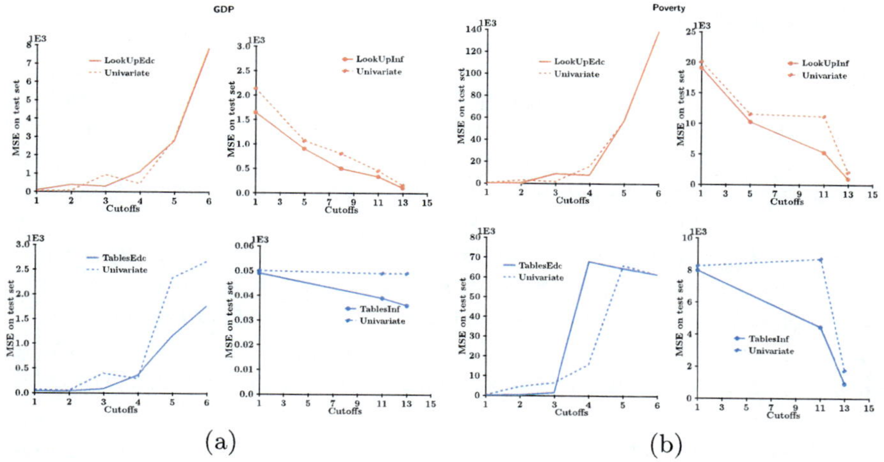


**Fig. 5** A comparison of testing error for models built from different versions of the system against testing error of models built by using the univariate feature selection technique on the CIA factbook dataset. These plots show that most versions of our system produced more accurate models than *Univariate*, supporting the claim that using the knowledge graph yields a better selected feature set than using mere correlations in the data. The *TablesInf* variant system produces the best models for both problem domains.

number of selected columns increases when $\gamma_e$ increases and $\gamma_i$ decreases. For each cutoff, linear models were built using linear regression with 10-fold cross validation, in which 80% of data were used as a training set, and an average mean square error (MSE) for each model computed on a test set for comparison purposes.

The results summarized in Fig. 5 clearly show overfitting when many columns are used to build models. When comparing at the same cutoff, our system tends to produce more accurate models than *Univariate*, indicating that model generalization is improved and the features selected using the knowledge graph are superior to features selected by correlations in the data. When comparing the accuracy of models produced from different versions of our system, using the Wikipedia tables method produces lower error models than using DBpedia lookup service. Since the former returns more accurate concepts than the latter, the relevance score calculated by the inference engine is more accurate for each feature producing a better group of features to build the model (comparing for the same cutoff, the number of features selected is smaller). Also, *TablesInf* built the best models. Since the Wikipedia tables method tends to return more specific concepts (most of them are leaves of the taxonomic hierarchy) than the DBpedia lookup service, it better suits with the information theoretic technique in aiming to find the most specific categories that subsume the particular concepts.

## 6 Related Work

The number of features is a key factor that determines the size of the hypothesis space, which contains all hypotheses that can be learned from the data. The larger the hypothesis space, the more potential models a learning method must eliminate. Many feature selection techniques have been developed to shrink the hypothesis space by removing irrelevant or redundant features, helping constrain the search for hypotheses. Typically feature selection is performed in one of two ways [9]: one is to rank features according to some criterion and select the top-ranked features [10–12], and the other is to select a minimum subset of features without deteriorating the learning performance [13, 14]. Our work fits within the feature ranking approach, where the criteria we use to rank features is derived from knowledge in a knowledge graph.

To further address a broad range of hurdles, such as the lack of quality training data and poor interpretability of machine learning [15], Knowledge-Infused Learning has been recently proposed as a new paradigm, in which it is the blending of learning methods with structured knowledge, especially the knowledge graphs [16, 17]. The main challenge of such an approach, however, is the impedance mismatch due to different representational forms and abstractions between symbolic and statistical AI techniques [15]. In order to resolve this challenge, three stages for infusion of knowledge into the machine learning process, especially deep learning architectures, have been developed: shallow [18–20] semi-deep [16, 21], and deep, regarding the form of infused knowledge and the steps of learning process in which this knowledge is infused. The proposed framework in this paper is considered as the shallow infusion, in which correlations from two sources are sequentially combined to build an accurate model, as it imposes a separation of these two factors: measures from the knowledge graph are used, first, to constrain a hypothesis space and then correlations within the data may be used to find the best model from that space. The goal of this work, however, is to develop a new end-to-end learning process that incorporates structural information encoded in a knowledge graph to perform feature selection prior learning a model, thereby it can broadly apply to various machine learning techniques rather than focusing on a particular one.

## 7 Conclusion

The lessons learned from the preceding evaluation are that using features which correspond with concepts that are close (i.e., have low edge count or high information content) to the concept corresponding to the output feature produces more accurate models. This provides support for the claim that structural information in a knowledge graph expresses correlations that are differ from those appear in data themselves. After all, some correlations appearing in the data may be the consequence of noise.

As of future work, we aim to test the system with various datasets and then make it available online or develop it as a plug-in on web browsers or spreadsheet applications, so that a wide range of users can build a model for various phenomena of interest, thereby the valuable feedback may be provided for further improve the system. Moreover, we are going to make use of other knowledge graphs that encoded either general or domain-specific knowledge and exploit different structural information and relations within a given knowledge graph, which may be useful for determining features' relevance. Such additional improvements not only indicate the generality of the proposed model building framework but also improve the accuracy in evaluating the relevance of the given features.

## References

1. Smith G, Shah E (2002) Data dredging, bias, and confounding: they can all get you into BMJ and the friday papers. BMJ 325(7378):1437–1438
2. Dhar V (2013) Data Science and Prediction. CACM 56(12):64–73
3. Dougherty, E.: Feature-selection overfitting with small-sample classifier design. IEEE Intell. Syst. **20**(6) (2005)
4. Bizer C et al (2009) Dbpedia- a crystallization point for the web of data. Web Semant.: Sci. Serv. Agents World Wide Web 7:154–165
5. Bhagavatula, C., Noraset, T., Downey, D.: Methods for exploring and mining tables on Wikipedia. In: Proceedings of IDEA (2013)
6. Lee J, Kim M, Lee Y (1993) Information retrieval based on conceptual distance in IS-A hierarchies. J. Documentation 49(2):188–207
7. Rada R, Mili H, Bicknell E, Blettner M (1989) Development and application of a metric on semantic nets. IEEE Tran. Syst. Man. Cybern. 19(1):17–30
8. Resnik P (1999) Semantic similarity in a taxonomy: an information-based measure and its application to problems of ambiguity in natural language. J. Artif. Intell. Res. 11:95–130
9. Liu H, Motoda H (2007) Computational Methods of Feature Selection. Chapman and Hall/CRC, Boca Raton
10. Hong, S.: Use of contextual information for feature ranking and discretization. IEEE Trans. Knowl. Data Eng. **9**(5), 718–730 (1997)
11. Kira, K., Rendell, L.: A practical approach to feature selection. In: ICML, pp. 249–256 (1992)
12. Ruiz, R., Riquelme, J., Aguilar-Ruiz, J.: Fast feature ranking algorithm. In: Palade, V., Howlett, R.J., Jain, L. (eds) Knowledge-Based Intelligent Information and Engineering Systems. KES 2003. Lecture Notes in Computer Science, vol. 2773, pp. 325–331. Springer, Berlin (2003). https://doi.org/10.1007/978-3-540-45224-9_46
13. Kohavi R, John G (1997) Wrappers for feature subset selection. Artif Intell 97(1–2):273–324
14. Koller, D., Sahami, M.: Toward optimal feature selection. In: Proceedings of 13th ICML, pp. 284–292 (1996)
15. Sheth A, Gaur M, Kursuncu U, Wickramarachchi R (2019) Shades of knowledge-infused learning for enhancing deep learning. IEEE Internet Comput 23(6):54–63
16. Yang, B., Mitchell, T.: Leveraging knowledge bases in LSTMs for improving machine reading. In: Proceedings 55th Annual Meeting Association Computer Linguistics, pp. 1436–1446 (2017)
17. Kursuncu, U., Gaur, M., Sheth, A.: Knowledge infused learning (K-IL): towards deep incorporation of knowledge in deep learning. In: Proceedings AAAI Spring Symposium (2020)
18. Xu, J., Zhang, Z., Friedman, T., Liang, Y., Van den Broeck, G.: A semantic loss function for deep learning with symbolic knowledge. In: Proceedings of the ICML (2018)

19. Yang, Z., Dai, Z., Yang, Y., Carbonell, J., Salakhutdinov, R., Le, Q.: XLNet: Generalized Autoregressive Pretraining for Language Understanding. arXiv:1906.08237 (2019)
20. Peters, M., et al.: Deep contextualized word representations. In: Proceedings of the NAACL (2018)
21. Yi, K., Jian, Z., Chen, S., Zheng, N.: Feature Selective Small Object Detection via Knowledge-based Recurrent Attentive Neural Network. arXiv: 1803.05263 (2019)

# Study of Feature Selection for Gold Prices Forecasting Using Machine Learning Approach

**Wilawan Yathongkhum, Yongyut Laosiritaworn, Jakramate Bootkrajang, and Jeerayut Chaijaruwanich**

**Abstract** Gold has been used as a hedge in times of extreme economic uncertainty and financial instability. Due to the multi-factors that influence the gold market, it's hard to definitively determine which factors should be used to forecast the price of gold. Therefore, this study aims to investigate the feature selection methods for determining the influencing factors for the gold price forecasting model based on the machine learning approach. The factors related to gold prices including macroeconomic indicators, market data, and investor fear indicators were investigated through six different features selection methods: correlation coefficient (R), recursive feature elimination (RFE), stepwise feature elimination (SFE), sequential feature selection (SFS), decision tree (DT) and LASSO. To forecast gold price based on these factors, we have applied four different methods: ordinary least squares regression (OLS), Prophet, long short-term memory networks (LSTM), and multi-layer perceptron (MLP) on the validation dataset for daily periods. The experimental results indicated that the SFS method provides the optimal predictor variables. Furthermore, the forecasting models using the OLS method are suitable and outperform the others method. According to this study, the feature selection methods based on the machine learning approach can identify the influencing factors similar to the econometric methodology from various research. Consequently, applying the optimal predictors for constructing gold price forecasting models may contribute to assessing the financial risk which is beneficial for investors to make decisions and manage their portfolios.

---

W. Yathongkhum · J. Bootkrajang · J. Chaijaruwanich (✉)
Department of Computer Science, Faculty of Science, Chiang Mai University, Chiang Mai, Thailand
e-mail: jeerayut.c@cmu.ac.th

W. Yathongkhum
e-mail: wilawan_y@cmu.ac.th

J. Bootkrajang
e-mail: jakramate.b@cmu.ac.th

Y. Laosiritaworn
Department of Physics and Materials Science, Faculty of Science, Chiang Mai University, Chiang Mai, Thailand
e-mail: yongyut.laosiri@cmu.ac.th



P. Meesad et al. (eds.), *Proceedings of the 19th International Conference on Computing and Information Technology (IC2IT 2023)*, Lecture Notes in Networks and Systems 679,
https://doi.org/10.1007/978-3-031-30474-3_8

**Keywords** Gold Prices Forecasting · Feature Selection · Gold Prices Influencing Factors · Machine Learning

## 1 Introduction

Generally, gold has been used by domestic authorities and investors as a hedge or a safe haven in times of financial instability and extreme economic uncertainty. According to Mansor H. Ibrahim [1], they found gold is a potential benefit of investment during periods of consecutive stock market decline. Supporting this finding, Najib Shrydeh et al. [2] measure the effectiveness of hedging in major US financial markets by comparing stocks and gold. The results indicated that prior shocks in gold returns can explain the present volatility of stock returns and the hedging effectiveness of gold tends to decrease as stock market capitalization increases. Moreover, Ciner et al. [3] studied the relationships between stock markets, bonds, gold, currency, and oil-based on daily data from the US and UK. This research confirms that gold has been used as a monetary asset in both the US and the UK when the exchange rates drop significantly. In addition, the research from Beckmann et al. [4] confirmed the relationship between the gold price and currency depreciation, the gold price tends to increase when a dollar falls, which means increasing volatility of dollar exchange rates leads to robust hedging of gold prices. Similarly, the study of Giannellis and Koukouritakis [5] pointed out that domestic authorities such as central banks and governments used gold as a hedge to prevent high misalignment in their currencies exchange rates, particularly in a time of financial instability and economic uncertainty. Additionally, the results from Narinder Pal Singh and Navneet Joshi [6] stated that there is only long-run causality between gold prices and inflation which concluded that the investment in gold can be used as a hedging tool against inflation.

The trend of gold price change can help reduce the impact of unexpected market fluctuation and has benefits for the assessment of financial risk. Even though the gold market is complex, volatile, and influenced by various factors, many studies are exploring the factors which may explain the variation in gold prices. Rohan et al. [7] study the impact of the macroeconomic news releases on silver and gold prices and found a strongly responds between gold and Consumer Price Index (CPI), and gold also has a significant response to the unemployment rate, Producer Price Index (PPI) and Gross domestic product (GDP). Furthermore, Cengiz et al. [8] study the factors that affect gold prices and found that gold prices have the highest negative correlation with the US exchange rate and a positive correlation with oil prices. Referring to Mehmet Bilgin et al. [9] stated that, the positive shock in the real effective exchange rate and the global economic policy uncertainty index affected the price of gold. The negative change in the volatility index affects the positive response of gold prices, and the price of gold is sensitive to decreasing oil prices. Additionally, the study by Tianqi Jiang [10] showed the U.S. deficit, international political events, and the 10-year treasury inflation-indexed security (TIPS) are statistically significant to predict the price of gold. Consequently, due to the nonlinear feature of gold price and many factors that influence the gold market, making accurate predictions of the gold price is a very challenging task.

During the last decade, many researchers have tried to implement predictive models to forecast either future gold prices or the pattern of price fluctuation. Traditional mathematical models such as autoregressive integrated moving average (ARIMA) have been developed as a tool for gold price forecasting. Moreover, machine learning methods such as linear regression (LR), artificial neural networks (ANN) including multi-layer perceptron (MLP), and deep learning models such as long short-term memory networks (LSTM) and convolutional neural networks (CNN) have also been applied in time-series forecasting problems. According to Iftikhar Sami and Khurum Nazir [11] predicting future gold rates based on the price of precious metals, oil price, major stock index, EURUSD exchange rates, economic indicators in the USA, and the stock price of the major gold producers' companies. The result indicated that the stock value of a major company influences the gold rates more than the US economy. Although the performance of the LR model is lower than the ANN model, the difference is not significant and LR is faster than ANN in the training phase. Furthermore, A.R. Hatamlou and M. Deljavan [12] forecast the price of gold based on the economic variables of Iran and the USA, including SPDR gold shares, oil prices, and stock prices. They used LR, ARIMA, and MLP models for forecasting, the study stated that MLP outperforms all other methods and the most important factors affecting Iran's gold price are Tepix, the US Government's debt, oil price, global gold price, and economic variables of Iran such as consumer index, inflation, bank interest rate, and GDP. In addition, Khani et al. [13] forecasted gold prices using deep learning methods based on the variables including COVID-19 daily time-series data and the top 10 tickers on the stock market which have the highest values for each sector. After developing prediction models based on the variables which have a high correlation coefficient value, the results demonstrate that LSTM is a powerful prediction method compared with other methods. However, influence factors and effective modeling are difficult to specify from the various factors and the different forecasting models which have been applied in many researches.

It is noteworthy that the most used feature selection methods that have been applied before constructing the forecasting model in existing studies usually focus on a correlation-based method. Nevertheless, the correlation only represents how strongly one feature implies the other without consideration of correlation among features. The observations by Wilcox [14] stated that the presence of outliers and non-linear associations can produce small or insignificant correlation coefficients, which will cause the highly predictive features or otherwise relevant features to be discarded. Additionally, Wang et.al [15] studied correlation-based methods of feature selection and found that these methods do not work well for time series data. Therefore, the correlation-based method may not confirm the effectiveness of the selected factors. Furthermore, the challenges due to the multi-factors can influence the price of gold, the results from different research showed that the relationships between these factors and gold price are not apparent and also had a mixed reaction. Then, it is not possible to definitively determine which factors should be used to forecast the price of gold. This problem is the motivation for investigating the feature selection methods for identifying the factors which affect the price of gold based on the machine learning approach. This research contributes to the literature by attempting to gain insights on the feature selection methods which can determine the influencing factors to construct the predictive models and offer a more comprehensive understanding of the selected features.

## 2 Data and Variables

This research mentions the factors that are closely related to gold prices including macroeconomic indicators, market data, and investor fear indicators from the United States, and collected them from the period of January 1st, 2009 to June 30th, 2021. The gold price data and most variables i.e. the consumer price index (CPI), real effective exchange rate (REER), unemployment rate (UnR), the 10-year breakeven inflation rate (IR10), treasury inflation-protected securities (TIPS), crude oil price (WTI), the skewness (SKEW), volatility (VIX), and economic policy uncertainty (EPU) were collected from the Fed.[1] Moreover, the producer price index[2] (PPI), the SPDR gold shares,[3] the Dow Jones index,[4] and the financial stress index (FSI[5]) were retrieved from several sources. The variables description is shown in Table 1. However, the time series data must be manipulated before it can be analyzed. Therefore, the historical data were aggregated by combining multiple pieces of data into a set along with the specific time frame. First, we filter raw data by removing none or empty value data. Then, we divide the time series data into regular time intervals (alignment period) and compute a single value for the points in the alignment period by an average, or filtering the empty data with the previous value.

## 3 Feature Selection Methods

Feature selection is an essential process that reduces the number of features by selecting a subset of the most significant and relevant features for constructing a sensible model with better prediction power. The feature selection methods that focus on this research can be grouped into three categories including filter, wrapper, and embedded methods.

The correlation coefficient (R) is one of the filter methods which is a statistical measure of the similarity between two features, the selection process is done by selecting the variable which has an absolute correlation coefficient value higher

[1] Federal Reserve Bank of St. Louis: https://fred.stlouisfed.org/.

[2] PPI data: https://data.oecd.org/price/producer-price-indices-ppi.htm.

[3] SPDR gold shares data: http://www.spdrgoldshares.com/usa/historical-data/.

[4] Dow Jones index: https://finance.yahoo.com/quote/DJI/.

[5] Office of Financial Research: https://www.financialresearch.gov/financial-stress-index/.

**Table 1** The variables' description

| Variables | Description | Frequency |
|---|---|---|
| Gold | Fixing Price 10:30 a.m. in London Bullion Market | daily |
| Macroeconomic Indicators | | |
| IR10 | 10-Year Breakeven Inflation Rate | daily |
| TIPS | 10-Year Treasury Inflation-Indexed Security | daily |
| UnR | Unemployment Rate | monthly |
| CPI | Consumer Price Index: Total All Items | monthly |
| PPI | Producer price indices | monthly |
| REER | Real Broad Effective Exchange Rate | monthly |
| Market Data | | |
| SPDR | SPDR Gold Shares (GLD Close) | daily |
| WTI | Crude Oil Prices: West Texas Intermediate | daily |
| Dow | Dow Jones Industrial Average (Adj Close) | daily |
| Investor Fear Indicators | | |
| EPU | Economic Policy Uncertainty Index | daily |
| SKEW | CBOE SKEW Index | daily |
| FSI | OFR Financial Stress Index | daily |
| VIX | CBOE Volatility Index | daily |

than the threshold value. Moreover, the wrapper methods are consisting of recursive feature elimination (RFE) [16], stepwise feature elimination (SFE) [17], and sequential feature selection (SFS) [18]. The RFE requires a specified number of features for finding the subset of the highest predictive features based on a machine learning algorithm, the procedure starting with all features and removing the least important features until a specified number of features remained. In addition, the selection process of the SFE method starts with no features and allows features to enter the regression model one by one. The regressor which produces the smallest p-value less than the threshold value when testing the significance of the model was added to the selected feature set. Furthermore, the SFS procedure starts with no features, once the best single feature is selected, the procedure repeats by adding a new feature that maximizes a cross-validated score to the set of selected features until the specified number of features is reached. According to the embedded methods, this research is interested in the decision tree (DT) and the least absolute shrinkage and selection operator (LASSO) [17]. The DT method creates a regression model based on several input variables in the form of a tree structure. The constructing process involves calculating the best predictive feature and the features which have an importance value greater than the specified threshold were selected. LASSO is a modification of linear regression which is used over regression methods and modified the loss function during training to minimize the complexity of the model. The minimizing process is done by adding a penalty to the sum of the absolute coefficient values, reducing the coefficients of the less important features to zero, and the subset selection relies on the exact set of non-zero coefficients.

## 4 Gold Price Forecasting Models

This research approach demonstrates the capability of machine learning and deep learning methods on multivariate time series forecasting problems. The gold price forecasting models were constructed based on four different methods composed of the ordinary least squares (OLS), Prophet [19], multi-layer perceptron (MLP), and long short-term memory networks (LSTM) [20]. The OLS is a statistical method that is used for modeling to estimate the linear relationships between independent variables and a dependent variable, it is achieved by minimizing the residual sum of squares between the observed and predicted value. In addition, the Prophet is a generalized additive model that is decomposable into three main components: trend or growth function, seasonality function, and holiday function, it offers a flexible regression model and provides uncertainty options to control the prediction uncertainty interval. Moreover, MLP is a supervised learning algorithm that is suitable for none linearly separable problems such as regression prediction. MLP sample network comprises one or more layers of neurons, the input layer receives the input to be processed, the predictions are made on the output layer, and there may be one or more hidden layers between the input and output layers. Furthermore, the LSTM has the ability to learn long-term dependencies by storing useful information on memory cells and vanishing unnecessary information, and capable of extracting the pattern of the input sequences or time-series data. The LSTM architectures used in this research consist of four layers, the first one is the input layer, the second is the LSTM layer, the third layer is a fully connected layer, and the last one is a regression output layer.

## 5 Experiment and Discussion

In this research, the experiments have been implemented in Python 3.7, Statsmodels[6] has been used for statistical data exploration, and some statistical modeling and Scikit-learn[7] have been used for some feature selection and loss evaluations. Furthermore, Mlxtend[8] has been used for some feature selection, Prophet[9] has been used for modeling to perform the analysis and Keras[10] has been used for deep learning implementations. In addition, the effectiveness of the forecasting model can be measured

6 https://www.statsmodels.org/.

7 https://scikit-learn.org/stable/.

8 https://github.com/rasbt/mlxtend.

9 https://github.com/facebook/prophet.

10 https://keras.io/.

using the coefficient of determination ($R^2$), root mean squared error (RMSE), and mean absolute error (MAE). According to a huge difference in price changes between current prices and a few years earlier, the dataset was decided to split sequentially instead of random sampling. Therefore, the training data is the historical data from the period of 2009–2020, and the testing data are from the period of 1 January 2021 – 30 June 2021.

We obtain the ranking number of important features from various feature selection methods which provide the relevant and optimized features for constructing the forecasting model as shown in Table 2. The set of predictive variables which produce the smallest RMSE value will be noted. The results present that all variables are significant when using the RFE as a selection method. Moreover, the DT and R methods perform the optimal predictor variables set by denying the CPI variable, while the rest of other methods provide a set of nine different variables. The predictive variables set can be described in Table 3.

**Table 2** The ranking number of important features for data in a range of daily frequency

| Variables | Feature Selection Methods | | | | | |
|---|---|---|---|---|---|---|
| | R | RFE | SFE | SFS | LASSO | DT |
| IR10 | 6* | 1* | 9* | 8* | 7* | 7* |
| TIPS | 2* | 4* | 6* | 5* | 4* | 6* |
| UnR | 8* | 9* | 5* | 12 | 9* | 4* |
| CPI | 13 | 6* | 11 | 11 | 11 | 13 |
| PPI | 3* | 7* | 3* | 9* | 13 | 9* |
| REER | 10* | 5* | 10* | 10* | 6* | 3* |
| SPDR | 1* | 2* | 1* | 1* | 1* | 1* |
| WTI | 9* | 10* | 2* | 3* | 8* | 2* |
| Dow | 5* | 13* | 4* | 2* | 2* | 5* |
| EPU | 4* | 12* | 12 | 4* | 10* | 11* |
| SKEW | 12* | 11* | 13 | 13 | 12 | 12* |
| FSI | 7* | 3* | 8* | 7* | 5* | 8* |
| VIX | 11* | 8* | 7* | 6* | 3* | 10* |

* indicate the optimal predictor variables

**Table 3** The set of predictive variables for daily basis data

| | Method | Variables |
|---|---|---|
| F1 | DT and R | IR10, TIPS, UnR, PPI, REER, SPDR, WTI, Dow, EPU, SKEW, FSI, VIX |
| F2 | RFE | IR10, TIPS, UnR, CPI, PPI, REER, SPDR, WTI, Dow, EPU, SKEW, FSI, VIX |
| F3 | SFE | IR10, TIPS, UnR, PPI, REER, SPDR, WTI, Dow, FSI, VIX |
| F4 | SFS | IR10, TIPS, PPI, REER, SPDR, WTI, Dow, EPU, FSI, VIX |
| F5 | LASSO | IR10, TIPS, UnR, REER, SPDR, WTI, Dow, EPU, FSI, VIX |

**Table 4** The performance of gold price forecasting for tomorrow

| | OLS | | | Prophet | | | LSTM | | | MLP | | |
|---|---|---|---|---|---|---|---|---|---|---|---|---|
| | RMSE | MAE | $R^2$ | RMSE | MAE | $R^2$ | RMSE | MAE | $R^2$ | RMSE | MAE | $R^2$ |
| F1 | 8.96 | 6.95 | 0.98 | 10.64 | 8.72 | 0.97 | 11.29 | 8.54 | 0.97 | 9.76 | 7.59 | 0.98 |
| F2 | 9.08 | 7.10 | 0.98 | 10.58 | 8.67 | 0.97 | 9.94 | 7.88 | 0.97 | 10.17 | 8.00 | 0.97 |
| F3 | 9.03 | 6.99 | 0.98 | 10.38 | 8.45 | 0.97 | 11.00 | 8.93 | 0.97 | 9.33 | 7.30 | 0.98 |
| F4 | 8.92 | 6.92 | 0.98 | 10.52 | 8.61 | 0.97 | 11.00 | 8.93 | 0.97 | 9.16 | 7.21 | 0.98 |
| F5 | 9.02 | 6.97 | 0.98 | 10.84 | 8.92 | 0.97 | 11.59 | 9.30 | 0.96 | 9.02 | 7.02 | 0.98 |

The performance of the forecasting models for tomorrow's price shown in Table 4 are encouraging and indicates the performance of all forecasting models from four different methods is quite productive. These r-squared values are greater than 0.96, meaning the independent variables can explain about 96 percent of the gold price change. To determine the effectiveness of the forecasting methods, although the performance of the OLS model is a bit different from MLP, the OLS has the advantage of lower the RMSE and MAE values. Therefore, OLS is very powerful at prediction and achieved the best results in comparison to Prophet, LSTM, and MLP methods.

According to the OLS forecasting results as shown in Table 4, the variables from the fourth feature set provide the best forecasting performance with the lowest RMSE and MAE value. Moreover, the fourth predictive variables set as shown in Table 3 is the selected features from the SFS method. Apparently, the SFS method is the effective feature selection method and the optimal predictor variables for gold price forecasting for tomorrow consist of IR10, TIPS, PPI, REER, SPDR, WTI, Dow, EPU, FSI, and VIX. Contrary to our results, many studies show the relationship between the CPI and gold price. However, our findings are similar to Narinder Pal Singh and Navneet Joshi [6], who indicated that in the short run, the CPI and gold price do not show any causality.

Interestingly, these predictive variables not only produce effective forecasting results but also provide some insight into potential connections to the gold price. According to the findings from many research [4, 5, 21] they found a relationship between the gold price and REER, when considering gold as a hedge asset in a time of unstable foreign exchange markets. A higher REER level implies the appreciation of the USD which is expected to decrease gold price. Moreover, WTI crude oil appears to have a significant positive impact and effect on the price of gold in both the positive and the negative changes in price [22, 23]. A higher level of oil price increases the price of gold. Alternatively, if oil prices dramatically fell, it would accelerate inflation and increase the price of gold. Furthermore, the study from Tianqi Jiang [10] showed that the percent change in TIPS has significant in predicting the prices of gold. Investment in gold is a favorable substitution when TIPS is low and the return on investment is decreased. Likewise, the result from Rohan Christie–David et.al [7] showed that the PPI announcements have a stronger relationship with gold, and the investors may take advantage of the short-term trading opportunities on gold related to the PPI releases. Additionally, co-movements between the gold price and the investor fear indicators such as EPU, VIX, and FSI have been investigated. In a time of the international financial crisis, the EPU has an impact on gold prices in many countries [24]. The higher EPU level will increase the demand for gold and lead to a rise in the price. Similarly, VIX had an economically significant effect on gold, the negative changes in VIX lead to a positive response in gold price [9]. The response to changes in volatility is consistent with gold being used as a safe haven to avoid risks in an uncertain environment. Although OFR Financial Stress Index (FSI) and 10-year breakeven inflation rate (IR10) have not been used in other studies, these variables were selected over several feature selection methods, which means it is an important feature and has predictive power. Similarly, Dow and SPDR also have not been widely used to study the impact on the price of gold. However, they were used as factors to predict the direction of gold and forecast gold prices in various research [12, 25, 26].

## 6 Conclusion

In this research, the feature selection methods including filters, wrappers, and embedded methods were explored with the purpose to indicate the optimal feature for constructing predictive models for forecasting gold prices, and the forecasting models were implemented based on four different methods. To measure the effectiveness of the forecast over daily data, the evaluation metrics including R-squared, RMSE, and MAE were applied. The experiments indicated that the predictive models using the OLS method are suitable and accurate for forecasting the gold price for tomorrow. Furthermore, after investigating the variables set obtained from various features selection methods, we found that the SFS method can provide the optimal predictor variables that consist of IR10, TIPS, PPI, REER, SPDR, WTI, Dow, EPU, FSI, and VIX.

This research was not an approach to study insight into the relationship between these variables. However, according to the manifold economics research on the factors affecting gold price, the results are resemblances to our study. Apparently, the feature selection methods based on the machine learning approach can determine the influencing factors with the insights understanding of the selected features, similar to the econometric methodology. The possibility to apply daily data for constructing gold price forecasting models based on the features obtained from an effective feature selection method may contribute to assessing the financial risk, it's beneficial for investors to make decisions and manage their portfolios.

However, the economic and financial indicators are not the only factors that may affect gold prices. Generally, markets react to external information such as unpredictable events and investors' behavior which are exhibiting sentiment in dimensions of fear, anxiety, worry, greed, and other emotion. These perceptions can be interpreted through text analysis from news and opinion. Therefore, an interesting topic for further research may be the effective forecasting model by aggregating the feature selection and heterogeneous information from text analysis.

## References


1. Ibrahim MH (2012) Financial market risk and gold investment in an emerging market: the case of Malaysia. Int J Islam Middle East Financ Manag 5(1):25–34
2. Shrydeh N, Shahateet M, Mohammad S, Sumadi M (2019) The hedging effectiveness of gold against US stocks in a post financial crisis era. Cogent Econ Finance 7:1698268
3. Ciner C, Gurdgiev C, Lucey BM (2013) Hedges and safe havens: an examination of stocks, bonds, gold, oil and exchange rates. Int Rev Financ Anal 29:202–211
4. Beckmann J, Czudaj R, Pilbeam K (2015) Causality and volatility patterns between gold prices and exchange rates. North Am J Econ Finance 34:292–300
5. Giannellis N, Koukouritakis M (2019) Gold price and exchange rates: a panel smooth transition regression model for the G7 countries. North Am J Econ Finance 49:27–46
6. Singh NP, Joshi N (2019) Investigating gold investment as an inflationary hedge. Bus Perspect Res 7(1):30–41
7. Christie-David R, Chaudhry M, Koch TW (2000) Do macroeconomic news releases affect gold and silver prices. J Econ Bus 52:405–421
8. Toraman C, Başarır C, Bayramoglu MF (2011) Determination of factors affecting the price of gold: a study of MGARCH mode Bus Econ Res J 2:37–50
9. Bilgin MH, Gozgor G, Lau CK, Sheng X (2018) The effects of uncertainty measures on the price of gold. Int Rev Financ Anal 58:1–7
10. Jiang T (2022) Economic variables affecting the gold price: the partial effects of linear models. In: 2020 2nd International conference on economic management and model engineering (ICEMME), pp 530–537. IEEE Xplore, Chongqing, 2020
11. Sami I, Nazir K (2018) Predicting future gold rates using machine learning approach. Int J Adv Comput Sci Appl 8:92–99
12. Hatamlou AR, Deljavan M (2019) Forecasting gold price using data mining techniques by considering new factors. J AI Data Min 7(3):411–420
13. Khani MM, Vahidnia S, Abbasi A (2021) A deep learning based methods for forecasting gold price with respect to pandemics. SN Comput Sci 2(4):335
14. Wilcox RR (2001) Modern insights about pearson's correlation and least squares regression. Int J Sel Assess 9(1&2):195–205
15. Wang QG, Li X, Qin Q, Huy NG (2013) Feature selection for time series modeling. J Intell Learn Syst Appl 5:152–164
16. Guyon I, Weston J, Barnhill S, Vapnik V (2002) Gene selection for cancer classification using support vector machines. Mach Learn 46(1):389–422
17. Pedregosa F et al (2011) Scikit-learn: machine learning in python. JMLR 12:2825–2830
18. Ferri FJ, Pudil P, Hatef M, Kittler J (1994) Comparative study of techniques for large-scale feature selection. In: Gelsema ES, Kanal LS (eds) Machine intelligence and pattern recognition, vol 16, pp 403–413. North-Holland, 1994
19. Taylor SJ, Letham B (2018) Forecasting at scale. PeerJ Preprints 37–45
20. Hochreiter S, Schmidhuber J (1997) Long short-term memory. Neural Comput 9(8):1735–1780
21. Reboredo JC (2013) Is gold a safe haven or a hedge for the US dollar? Implications for risk management. J Bank Finance 37:2665–2676
22. Gil-Alanaa LA, Yaya OS, Awe OO (2017) Time series analysis of co-movements in the prices of gold and oil: Fractional cointegration approach. Resour Policy 53:117–124
23. Reboredo JC (2013) Is gold a hedge or safe haven against oil price movements. Resour Policy 38(2):130–137

24. Gao C, Da-ming Y, Jin-yu C (2019) Dynamic response pattern of gold prices to economic policy uncertainty. Trans Nonferrous Met Soc China 29(12):2667–2676
25. Sadorsky P (2021) Predicting gold and silver price direction using tree-based classifiers. J. Risk Financ. Manag 14(5):198
26. Hafezi R, Akhavan AN (2018) Forecasting gold price changes: application of an equipped artificial neural network. AUT J Model Simul 50(1):71–82

# Airbnb Occupancy Rate Influential Detection Based on Hosting Descriptions with LDA

**Rattapon Choogortoud, Dittapol Muntham, Worawek Chuethong, Sart Srisoontorn, Orasa limpaporn, and Maleerat Maliyaem**

**Abstract** At present, the Airbnb rental business is constantly increasing, which is caused competition in the room rental business. This study proposed an alternative of writing host descriptions for rentals to find hidden topics using Latent Dirichlet Allocation (LDA) as Natural Language Processing (NLP) technique in order to help owners increase their occupancy rate. Airbnb booking data was taken from the website only for Bangkok's Airbnb area, including 17,074 rooms between June 2022 and June 2023. The study found that having a host description resulted in a booking average of over 8.47%, with the top five most effective topics being room function, activity, transportation, facility, and travel, respectively. The host's description is a choice that owner shows tenant the features of the room they can choose for booking.

**Keywords** Machine Learning · Natural Language Processing · Latent Dirichlet Allocation · Airbnb

---

R. Choogortoud · D. Muntham · W. Chuethong · S. Srisoontorn · O. limpaporn · M. Maliyaem (✉)
Faculty of Information Technology and Digital Innovation, King Mongkut's University of Technology North Bangkok, Bangkok, Thailand
e-mail: maleerat.m@itd.kmutnb.ac.th

R. Choogortoud
e-mail: rattapon.c@email.kmutnb.ac.th

W. Chuethong
e-mail: chuethong.w@email.kmutnb.ac.th

S. Srisoontorn
e-mail: sart.s@email.kmutnb.ac.th

O. limpaporn
e-mail: orasa.l@email.kmutnb.ac.th



P. Meesad et al. (eds.), *Proceedings of the 19th International Conference on Computing and Information Technology (IC2IT 2023)*, Lecture Notes in Networks and Systems 679,
https://doi.org/10.1007/978-3-031-30474-3_9

# 1 Introduction

At present, there are a lot of vacant rooms available, such as apartments, condominiums and houses, especially in large urban tourism areas. The opening of rental accommodations is beneficial for boosting the local population's economy and tourism. There are distinctive selling qualities for each city and nation, whether in the accommodations' features, past visitors' pleasures, locations, etc. Bangkok is the capital of Thailand and the most visited city in Thailand [1]. For these reasons, the Airbnb rental business continues to grow, making it very competitive in renting. The host description given by the owner is one of the things that encourages travelers to book a room [2]. Booking decisions by travelers may be influenced by the effect of the host description on the occupancy rate. The Latent Dirichlet Allocation (LDA) technique from Natural Language Processing (NLP) is a popular method for extracting hidden topics from descriptions [3, 4]. This study's goal is to identify the hidden topics in the host description that will raise occupancy rates.

# 2 Literature Review

Airbnb is an online platform where people can search for renting rooms or accommodations during their vacation. Typically, these areas consist of either a "whole location" (house, condominium, etc.) or a "private room" in a home where the host is also present. For "guests", utilizing Airbnb is generally similar to using an internet travel agency to find and reserve lodging, however booking may necessitate direct contact with the "host". For hosts who own and/or manage Airbnb listings, the platform provides the opportunity to freely and practically effortlessly transform into a tourist lodging provider by posting descriptions and photos of one's lodging, corresponding with potential guests, and accepting bookings and payments from all over the world [5]. Airbnb is an innovative challenge to traditional hospitality that is referred to as the "Sharing Economy" or "Peer to Peer, P2P". Numerous significant and ongoing factors, including technology development, socio-cultural change, and globalization, contribute to the success of networked platforms like Airbnb. Certainly, in the foreseeable future, these factors will keep the phenomena alive [6].

Airbnb hosts can advertise their accommodational functions, locations, transportation, and others—called host descriptions—on the website to attract visitors. The host description is an opportunistic strategy to combine with various factors [2]. The majority of the satisfied subjects include aspects of the destination and the surrounding region (such as accommodation, location, amenities, shopping and restaurant in the area, etc.). In practice, these topics are a part of a guide for hosts to provide better information for their guests for booking decisions. [7] However, the host description has several topics to consider. Thus, only observation can cause you to miss some topics known as "hidden topics".

The Latent Dirichlet Allocation (LDA) technique is a popular topic modeling approach for identifying hidden topics. LDA has been widely applied to various fields, including natural language processing, text mining, and social media analysis, in order to understand the reactions and conversations between people in online communities [8]. In recent years, LDA has also been utilized for analyzing social media data, building recommender systems, and analyzing online reviews. The online platform of Airbnb, which comprises host descriptions and review comments, provides an opportunity for LDA to explore hidden topics and grow a large number of successful Airbnb businesses [2, 4, 7, 8]. However, there has been little research on the application of LDA to Airbnb host descriptions. Therefore, further research using LDA can help to uncover new insights and improve the effectiveness of Airbnb's business strategies.

## 3 Method

In this exploratory study, we provide a conceptual framework for the Airbnb owner's host description, as shown in Fig. 1.

### *3.1 Data Collection*

The dataset was gathered from the website name "Inside Airbnb" (http://insideairbnb.com). The raw data is a publicly accessible and non-commercial website. The Airbnb data set represents many dimensions, for listing id, host id, hosting descriptions and booking. However, this research focused on date hosting descriptions.

The Bangkok data set includes 17,074 rooms. There are 1395 rooms with Hosting descriptions in English, 2,874 rooms with multiple languages, and 905 rooms with no Hosting descriptions. Calendar data consisted of 6,231,645 lists that were conducted to find the occupancy rate of the rented room on Bangkok's Airbnb platform from June 20, 2022, to June 20, 2023.

### *3.2 Language Detection*

Before data analysis, data preprocessing is a crucial process in Language Detection (LD) tasks because there are many languages, such as English, Thai, and Chinese. The data were cleaned by removing noise and tidying the text because we were only interested in analyzing English host descriptions. LD is the processing in block A (See Fig. 1), where the host descriptions were detected using the ISO 639–1 language code and ISO 639–1 language name. Subsequently, the data were filtered to only be in English and recorded in data set A.

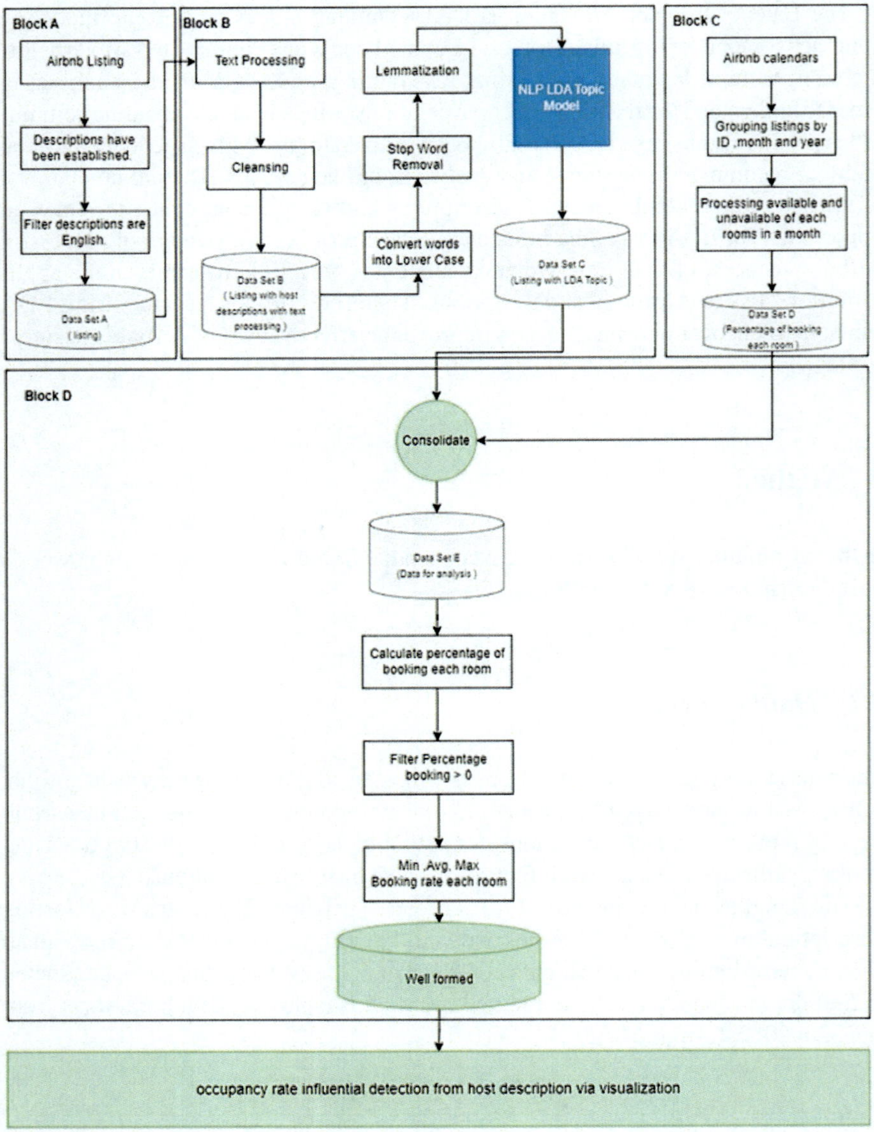


**Fig. 1** Occupancy Rate Influential Detections Framework.

## *3.3 Text Processing*

Text processing is the technique for flagging host descriptions as defined or undefined by converting words into lowercase, stopping words and lemmatization. The cleansing data was recorded in the data set B. The data set B was processed to find hidden topics before using Latent Dirichlet Allocation (LDA).

## 3.4 *Latent Dirichlet Allocation*

A generative probabilistic model of a corpus is called Latent Dirichlet Allocation (LDA). The fundamental concept is that the documents are viewed as random mixtures of latent topics, where a subject is represented by a distribution over words [8]. LDA, first introduced by Blei, Ng and Jordan in 2003 [9].

LDA assumes the following generative process for each document w in a corpus:

(a). Choose N—Poisson (ξ)
(b). Choose θ—Dir (α)
(c). For each of the N word wn

i. Choose a topic Zn—Multinomial (θ)
ii. Choose a word wn from P(wn|Zn, β), a multinomial probability conditioned on the topic

In Fig. 2, the LDA model is shown as a probabilistic graphical model. The LDA representation has three levels, as shown in the picture. The parameters α and β are corpus-level parameters that are presumably sampled once during the creation of a corpus. The variables θd are sampled once for each document and are document-level variables. Finally, the word-level variables zdn and wdn are sampled once for each word in each document [9]. In this study, the Coherence score was utilized to determine the appropriate number of topics for the LDA model.

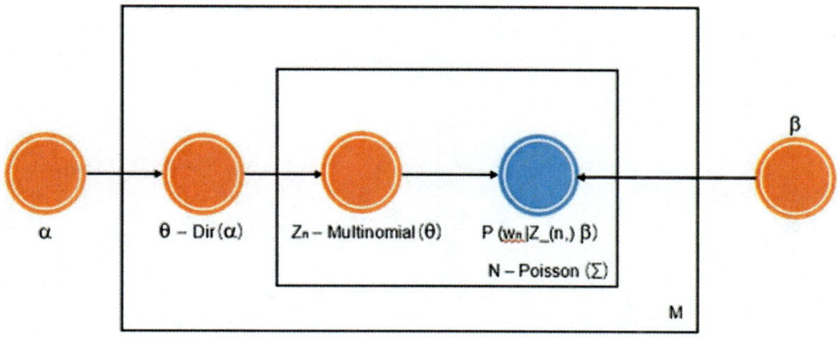


**Fig. 2** Overview of LDA system.

### 3.5 *Consolidate Data*

The data set D was the data from the booking Airbnb calendars that included ID, months and years. After that the data from the data set C and the data set D were gathered and recorded to the data set E (See Fig. 1). The occupancy rate influential detection from the data set E was performed by LDA modeling and visualization.

## 4 Result

The results of the analysis demonstrated that Airbnb listings with host descriptions had a higher booking rate compared to those without host descriptions (8.47%). Based on data from Bangkok's Airbnb platform from June 20, 2022 to June 20, 2023, the booking rate was calculated as the number of bookings per year (times 365 days) × 100.

The coherence score showed the highest score at 1 topic, which is a difficult topic to estimate. The coherence score decreased from 1 to 14 topics and increased at 15 topics to peak at 20 topics (See Fig. 3). Thus, we choose 20 topics (coherence score = 0.4682) for keyword grouping and priority arrangement with LDA model. LDA model showed 20 bubbles on the left-hand side plot (See Fig. 4) and bubbles were chosen from the individual bubble and more percentage than the booking average (12.40%).

Table 1 showed five categories of host descriptions: room function, activity, transportation, facility and travel, respectively. The room function (14.93%) was the highest average percentage of bookings, with 10 keywords such as "bed," "size"

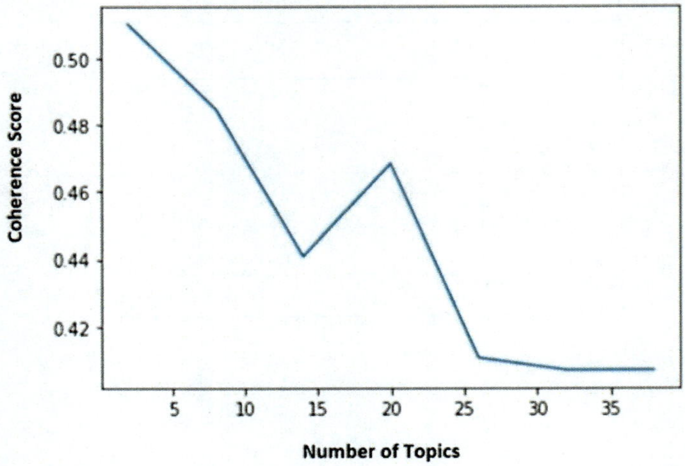


**Fig. 3** Feature Engineering to Optimization Number of Topic LDA.

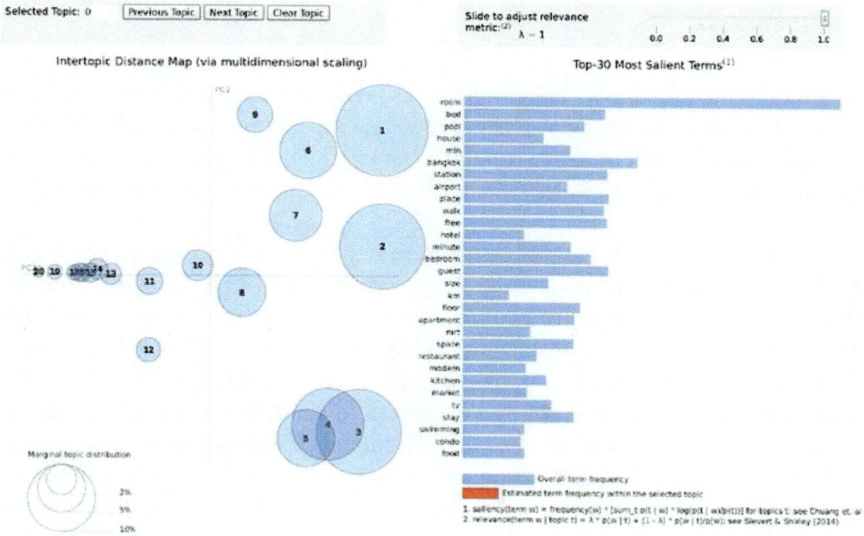


**Fig. 4** Host descriptions Topic from Latent Dirichlet Allocation model.

and "studio" that were the first priorities of the booking decision. The second topic was activity that included keywords about places such as "restaurant", "food", "bar", "shop", and "street".

**Table 1** Top 5 topics: topic label, keywords and average percentage of booking

| Topic | Topic Label | Keywords | Average percentage of booking |
|---|---|---|---|
| 1 | Room Function | bed, size, studio, king, queen, bathroom, sofa, table, tv, comfortable | 14.93 |
| 2 | Activity | restaurant, food, local, place, close, shop, street, nearby, bar, quiet | 13.82 |
| 3 | Transportation | station, walk, minute, mall, shopping, center, Sukhumvit, locate, central, just | 13.74 |
| 4 | Facility | bedroom, floor, apartment, space, area, fully, living, kitchen, view, room | 13.38 |
| 5 | Travel | km, patpong, reachable, donmuang, spot, smoke, prepare, tour, chang, patio | 12.50 |

## 5 Discussion

The study found that the host description was one of the factors that influenced the average percentage of booking. Each hidden topic in the host description affects a different occupancy rate, which is similar to Janssens [2]. The first five subjects from this study were similar to Keawtoomla [7], which found that the words expressed favorable opinions.

The result of this study showed host descriptions affected the average percentage of booking. The LDA technique had the potential to find hidden topics, although it had an average percentage of booking. However, it was limited and complicated because this technique was still manual to determine the topic labels of each topic. In addition, there are many variables affected on booking rate such as locations, room types and amenity. The words are a part of the host description that raised the booking rate. However, the success of an Airbnb business does not depend only on the host description. The hosting service (flexible check-in and check-out) and the customer reviews posted on that website are other important factors in the success of the Airbnb business [2]. It is possible to find a hidden topic in both host descriptions and review comments for long-term Airbnb business.

Using LDA for text mining with topic modeling is useful to understand how people respond to social media online, such as host descriptions, review comments, and other conversations [2, 4, 7]. Furthermore, LDA is beneficial for implementing keywords that help owners decide to write a host description to increase occupancy rates.

## 6 Discussion Limitation and Future Research

The study had limitations in that it focused only on English host descriptions and used only one city's data. Future research should be extended to Airbnb in other cities and host descriptions should be compiled in other languages. Additionally, the themes discovered using the LDA approach vary depending on the data. As a result, the ontology technique should be used in the next study to clarify the subject.

## References

1. Bessadeg F (2022) The world's most visited cities. https://travelness.com/most-visited-cities-in-the-world. Accessed 25 Oct 2022
2. Bram J, Matthias B, Van den Poel D (2021) Evaluating the influence of Airbnb listings' descriptions on demand. Int J Hospitality Manag 99(6):1–11. https://doi.org/10.1016/j.ijhm.2021.103071
3. Muhammad RR, Walayat H, Asaf V (2022) Performance analysis of deep approaches on airbnb sentiment reviews. In: 2022 10th International symposium on digital forensics and security (ISDFS), 2022. https://doi.org/10.1109/ISDFS55398.2022.9800816

4. Mohamed C, Omar B, Younes C (2021) Towards a machine learning and data mining approach to identify customer satisfaction factors on Airbnb. In: 2021 7th International conference on optimization and applications (ICOA), 2021. https://doi.org/10.1109/ICOA51614.2021.9442657
5. Daniel G (2019) Progress on Airbnb: a literature review. J Hosp Tour Technol 10(4):814–844. https://doi.org/10.1108/JHTT-08-2018-0075
6. Jeroen O, Albert B (2016) Airbnb: the future of networked hospitality businesses. J Hosp Tour Technol 2(1):22–42. https://doi.org/10.1108/JTF-11-2015-0048
7. Nathakit K, Nathakit K, Jakramate B (2022) Using latent Dirichlet allocation to investigate guest experience in Airbnb accommodation during COVID-19 pandemic in the United Kingdom. In: 2022 19th International joint conference on computer science and software engineering (JCSSE), 2022.https://doi.org/10.1109/JCSSE54890.2022.9836314
8. Hamed J, Yongli W, Chi Y, Xia F, Xiahui J, Yanchao L, Liang Z (2019) Latent dirichlet allocation (LDA) and topic modeling: models, applications, a survey. Multimedia Tools Appl 78:15169–15211. https://doi.org/10.1007/s11042-018-6894-4
9. David MB, Andrew YN, Michael IJP (2003) Latent Dirichlet allocation. J Mach Learn Res 3:993–1022

# Projectile Launch Point Prediction via Multiple LSTM Networks

**Wisit Wiputgasemsuk**

**Abstract** Launch point prediction of projectile targets is the main task of ground-based air surveillance radars in military services. Traditionally, the radar captures the portion of the projectile's trajectory to obtain time series data of position and velocity, and then uses an extrapolation method to determine the launch point. This work aims to enhance the accuracy of launch point prediction from the radar measurement data through the development of multiple Long-Short Term Memory (LSTM) networks. The proposed method consists of three stages. First, an LSTM-based filtering model is built to filter the noisy radar data and estimate the flight trajectory states. Second, an LSTM-based classifier model is created to identify the class of munitions based on their flight trajectories. Finally, multiple LSTM networks are developed to serve as the launch point predictor, in which each network is trained on different sets of munition classes. A switching law based on the classification result was established to select the best launch point predictor network for a specific class of munitions. The prediction results demonstrated the promising performance of the proposed method in launch point prediction. Furthermore, the correlation between the data distribution and the predictive performance of the network was explored.

**Keywords** Launch Point Prediction · Multiple Networks · Long-Short Term Memory

## 1 Introduction

Counter battery radars [1] play an important role in counter-fire operations. Their main function is to classify and forecast the launch point of hostile artillery, which typically includes rocket launchers, howitzers, and mortars. This facilitates the identification of the type of hostile artillery and the determination of its firing position through analysis of the flight trajectory of its shells. Radar is often used to capture the

W. Wiputgasemsuk (✉)
Aeronautical Engineering Division, Defence Technology Institute, Nonthaburi, Thailand
e-mail: wisit.w@dti.or.th



P. Meesad et al. (eds.), *Proceedings of the 19th International Conference on Computing and Information Technology (IC2IT 2023)*, Lecture Notes in Networks and Systems 679,
https://doi.org/10.1007/978-3-031-30474-3_10

portion of the flight trajectory of a projectile, allowing for the acquisition of time-history profiles of its position and velocity. Traditionally, to determine the launch point, the noisy radar measurement data is processed using a filtering technique to estimate the trajectory states of the projectile target. This information, along with a kinematic model of the projectile, is then used in a regression analysis to calculate the launch point [2–4].

In recent years, the advancement of neural networks has led to an increasing number of research studies that use this approach for the classification and launch point prediction of projectile targets. Carpenter et al. [5, 6] utilized the feed-forward neural network to identify the types of munitions, including 70 mm, 107 mm, and 122 mm, from radar measurement data. They also used the neural network as a functional representation instead of the piecewise solution from the traditional numerical calculation. Eckert et al. [7] employed a deep learning neural network to classify the variations of missiles. They built 12 distinct classes of missiles and added the simulated radar noise to each simulated trajectory within each class. Their results showed that the deep learning neural network achieved nearly 100% classification accuracy, while shallow neural networks did not perform well. Kim et al. [8] used an LSTM network for predicting the trajectory and launch point of ballistic objects from radar measurement data. The LSTM model was trained using true simulated trajectories. They used an Unscented Kalman Filter (UKF) to reduce measurement noise, and then used the filtered data for testing the performance of the LSTM model. Hou and Liu [9] utilized an LSTM network to estimate the real position of projectile trajectories from noisy radar measurements and developed a Mixture Density Network (MDN) for trajectory extrapolation and launch point prediction. They integrated the LSTM and MDN into a single end-to-end network, and trained it using radar measurement samples and their corresponding ground truth launch points. They also compared the traditional Extended Kalman Filter (EKF), a vanilla LSTM, and the LSTM + MDN for predictive performance.

In the literature, the use of neural network has been demonstrated to be effective in classifying and predicting the launch points of projectile targets. However, considering the munitions in the real world, they have various classes, each with unique flight trajectory characteristics. If the trajectory data for all classes were combined into a single training dataset, the resulting data would be highly diverse. Additionally, the radar measurement data is often noisy and restrictively available in practice, so training the network on the such dataset can lead to poor generalization performance. Indeed, some techniques, such as regularization, cross-validation, and early stopping callback, can help to improve the generalization. However, since the training data in this particular case is full of noise and contains outliers, more than those techniques might be required to achieve accuracy at an acceptable level. One such approach is to build separate networks for each class of munitions. By having prior knowledge that the characteristics of flight trajectories between classes are quite different, building several networks by training each on a certain class of munitions, could further improve the overall predictive abilities of the networks. This idea is not new. It was previously studied and analyzed in the literature. Jacops et al. [10] introduced a supervise learning procedure using the mixtures of local experts. Their idea is that if we know in advance that a set of training cases may be naturally divided into subsets that correspond to distinct subtasks, interference can be reduced by using a system composed of several different "expert" networks plus a gating network that decides which of the experts should be used for each training case. Nguyen and Chan [11] presented multiple neural networks for long term time series prediction, with each model forecasting at a different time horizon. They showed that using multiple networks could yield better result than using a single network, as it could help reduce the propagation errors that grew during recursive prediction for long-range forecasts.

To this end, this paper proposes the method of utilizing multiple LSTM networks to enhance the accuracy of launch point prediction. The launch point predictor model consists of multiple LSTM networks which are trained on different sets of trajectory data. The LSTM-based classifier is built to identify the class of munition from its flight trajectories. The switching law based on the classification result is established to select the best launch point predictor network for each class of munitions. The main contributions of this paper are:

- Introducing the use of multiple LSTM networks to forecast the launch points of projectile targets from noisy radar measurement data, and
- Using multiple LSTM networks plus a switching law can improve the accuracy of launch point predictions compared to using a single LSTM network.

## 2 Methodology

### *2.1 Proposed Method*

The basis of the proposed method is to make full use of LSTM's ability in sequence pattern recognition for processing, classifying and, predicting tasks. The framework of the proposed methodology is illustrated in Fig. 1. The method consists of three parts: the filter, the classifier, and the launch point predictor.

The first part of the proposed method is the filtering model. It is used to reduce the measurement noise and to estimate the real trajectory states. The LSTM network has shown effectiveness in filtering noise errors in measurement devices. For example, Jiang et al. [12] used the LSTM network to reduce random noise in MEMS IMU for navigation applications, which outperformed the conventional autoregressive moving average (ARMA) model. Hou et al. [9] used the LSTM network to predict the trajectory states of projectiles from noisy radar measurement data. In a similar way, this work uses the LSTM model to filter the noisy data in the first step. A segment of noisy radar data is fed into the filter model, which then outputs the estimated trajectory data of the same length as the input data.

The second part of the method is the classifier model, which predicts the munition class based on their trajectories. The estimated trajectory data from the filter is fed into the LSTM network, which then estimates the possibility to which of the class

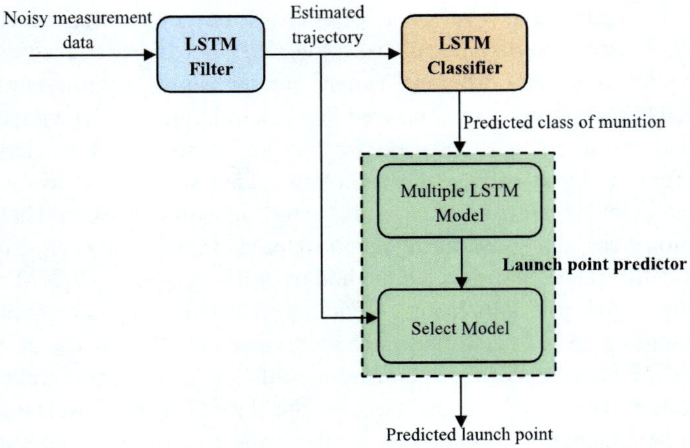


**Fig. 1** Workflow of the proposed method

that the trajectories belong. The output of the model is the most probable class of munitions.

The third part is the launch point predictor model, which consists of multiple LSTM networks, each trained on a different set of munition classes. The predicted munition class is input into the switching law to select the best model for launch point prediction. The estimated trajectory data is then fed into the selected model, and the output is the predicted launch point.

## 2.2 *Dataset*

The radar measurement data used in this study was generated from the trajectory simulation program in the fire control system [13]. This program used the kinematic model of artillery rockets to create trajectories by varying simulation parameters such as launch elevation, launch azimuth, initial spin rate, air density, wind speed, wind direction, and others. The variation of these parameters was distributed according to a normal distribution with specified standard deviations. The resulting trajectory data included three-dimensional positions (x, y, z) at the update rate of 5 Hz (0.2 s per data point).

Four different classes of artillery rockets were created by configuring differently in thrust and aerodynamic drag. These differences resulted in distinct down range and apogees. Assume that rockets were launched from nine different launch points, with the radar station set as the origin coordinate (0, 0). To provide a clearer illustration, some samples of the projectile trajectories were picked to show in Fig. 2 (left). Table 1 summarizes the statistics of the trajectories in the training dataset. For each class in the dataset, there were 4,500 trajectories for training and an additional set of 900 trajectories for both validation and testing. The training, validation and testing data were in the ratio of 72:14:14, respectively. This ratio was appropriate since the validation and testing data were substantial enough to evaluate the model's performance on unseen data. Actually, the choice of ratio between the training and validation data depends on the size of dataset and the complexity of the model. Determining the optimal ratio for a specific problem and dataset requires an empirical process to minimize both the validation error and the error rate on the training set [14].

In general, a radar system can maintain tracking of a targets as long as the characteristics of the target and the operational environment are within the capabilities of the radar system. However, there are various factors that may cause a radar system to lose tracking such as target being out of range, target moving at high speeds, or the presence of interference like jamming. To represent the partially available radar data, the entire trajectory was segmented into multiple windows. By having longer window lengths, the accuracy of launch point prediction has the potential to increase. Assuming that the radar could track a target for 10 s during the time interval from 25 to 37 s after launch, and with the update rate of 5 Hz, the resulting window had a length of 50 time steps, starting from time step 125 and ending at time step 185. Shifting the window by one time step resulted in 11 overlapping windows for each trajectory. To make the data more realistic, additive white Gaussian noise with a

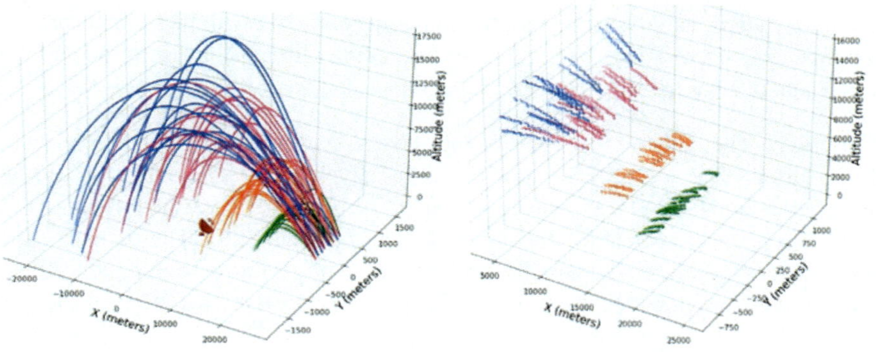


**Fig. 2** (left) Projectile trajectories and (right) portions of noisy radar data

**Table 1** Flight Statistics of 4 classes of rocket (***A***, ***B***, ***C***, ***D***)

| Classes | Trajectories | Down Range (meters) | | | Cross Range (meters) | | | Apogee (meters) | | |
|---|---|---|---|---|---|---|---|---|---|---|
| | | Mean | Min | Max | Mean | Min | Max | Mean | Min | Max |
| ***A*** | 4500 | 18,662 | 17,345 | 19,532 | −18 | −1909 | 1805 | 6512 | 3733 | 9453 |
| ***B*** | 4500 | 10,729 | 9910 | 11,131 | 6 | −784 | 819 | 3379 | 1917 | 4794 |
| ***C*** | 4500 | 41,290 | 31,584 | 50,587 | −21 | −4344 | 3968 | 13,668 | 7082 | 22,255 |
| ***D*** | 4500 | 33,301 | 28,193 | 39,725 | 17 | −3188 | 3687 | 11,476 | 6378 | 18,195 |

standard deviation of 10 m for each axis was applied to data points. Some examples of incomplete and noisy radar data are shown in Fig. 2 (right).

### 2.3 LSTM Networks and Training

All networks were implemented using the TensorFlow framework on a Windows10 system equipped with an Intel Xeon E3-1505 M 2.8 GHz CPU and 32 GB of RAM. Before training, the datasets were standardized using the StandardScaler function provided by Scikit-Learn. In the training process, the number of layers and neurons were tuned by random search and adjusted manually around the best value found by random search. At each time, the early stopping callback was used to avoid overfitting and to restore the best weights. The training process and hyperparameters for each network are as follows:

**Filter:** The network consisted of four layers, with the first being a flatten layer of 150 neurons (representing 50 time steps $\times$ 3 input features). The hidden layers were two LSTM layers with 28 and 20 neurons, respectively. The output layer was a dense layer with 150 neurons, matching the number of neurons in the first layer. In the training phase, the input sequence was the noisy trajectory data (x, y, z) with a length of 50 time steps, and the target was the true trajectory data with the same length as the input sequence. The Adam optimizer was used, with a batch size of 50 and the mean square error (MSE) as the loss function.

**Classifier:** The network consisted of four layers, starting with a flatten layer with 150 neurons. The second layer was an LSTM layer with 4 neurons, followed by a dense layer with 4 ReLU activation functions. The final layer had 4 cells with a softmax activation function to output the class probabilities. The output of the network was the class label with the highest probability. During training, the Adam optimizer was used, with a batch size of 100 and categorical cross entropy as the loss function. Accuracy was used as the evaluation metric.

**Launch Point Predictor:** To find the best launch point predictor for each class of munition, seven alternative LSTM networks were developed and trained on a different dataset, as outlined in Table 2. All alternative networks had the same architecture but differed in the number of neurons in the hidden LSTM layers. The first layer was a flatten layer with 150 neurons, followed by three LSTM layers with varying numbers of neurons, as listed in the third column of Table 2. The final layer was a dense layer with 2 neurons corresponding to the target (x, y). In the training phase, the Adam optimizer was used, with a batch size of 50 and the MSE as the loss function. The input sequence for training was the true trajectory data over 50 time steps, and the target was the corresponding actual launch points (x, y). The MSE losses for training and validation were listed in the last two columns of Table 2. We can observe that the MSE value for validation was slightly lower than the value for training. This was

**Table 2** The network architecture and the MSE for training/validation

| Networks | Datasets | Neurons | MSE | |
|---|---|---|---|---|
| | | | Training | Validation |
| *1* | *A* | 22,20,16 | 14.67e-6 | 13.08e-6 |
| *2* | *B* | 20,18,14 | 12.61e-6 | 5.46e-6 |
| *3* | *C* | 20,18,14 | 13.84e-6 | 8.68e-6 |
| *4* | *D* | 20,18,14 | 6.29e-6 | 2.19e-6 |
| *5* | $A + B$ | 22,20,16 | 20.46e-6 | 4.53e-6 |
| *6* | $C + D$ | 22,20,16 | 31.66e-6 | 5.30e-6 |
| *7* | $A + B + C + D$ | 22,20,16 | 9.95e-6 | 2.18e-6 |

due to the early stop callback that stopped the training process when the validation error reached its minimum and started to increase.

## 3 Experimental Results

To evaluate the effectiveness of the proposed method, a new set of 3,600 trajectories (900 trajectories per class) with added white Gaussian noise was generated to serve as the testing dataset. The same preprocessing steps were applied once for the testing dataset, which involved standardizing the data and dividing it into a window with a length of 50 time steps. The testing process started with feeding the segment of noisy data into the filter model, which then output the estimated trajectory. The root mean squared error (RMSE) was used as a metric to evaluate the deviation between the estimated and actual trajectories. The result showed the RMSE of 4.3 m across the entire testing dataset, as illustrated in Fig. 3, which showed the close proximity between the estimated and actual trajectories.

The estimated trajectories were then input into the classifier model. The result of the classification accuracy was found to be 100%, despite the network having a simple structure with few neurons. This was likely due to the clear separation between the trajectories of each class, making it easier for the network to classify accurately. After that, the estimated trajectories were fed into the alternative networks, as listed in Table 2. Once again, the RMSE was used as a metric to evaluate the performance of all alternative networks in comparison. The RMSE for launch point prediction is defined as follows:

$$\mathrm{RMSE} = \sqrt{\frac{1}{2n_{\mathrm{sample}}} \sum_{i=1}^{n_{\mathrm{sample}}} (x_i - \hat{x}_i)^2 + (y_i - \hat{y}_i)^2}$$

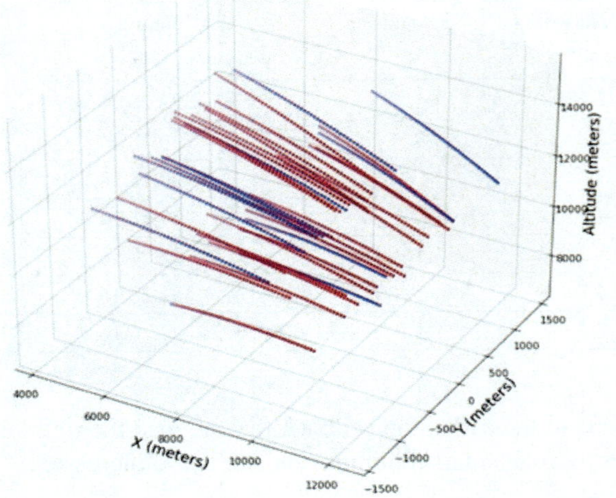


**Fig. 3** Comparison of the estimated trajectories (blue dot) and actual trajectories (red cross)

where $\hat{x}_i$, $\hat{y}_i$ represent the i-th predicted launch point, and $x_i$, $y_i$ represent the i-th actual launch point.

Table 3 compares the performance of all alternative networks on the launch point prediction. When comparing network *7*, which was trained on all combined data, with the individual networks ***1***, ***2***, ***3***, and ***4***, which were trained on individual datasets ***A***, ***B***, ***C***, and ***D***, respectively, it can be seen that all the individual networks except for network ***3*** perform better than network *7*. This result is consistent with the point of view that the data distribution in a specific dataset correlates with the network's predictive performance, regardless of how well the network is trained. Although all networks were comparably well-trained, network ***3*** which was trained on highly distributed ***C***-class data performs the worst, while network ***2*** which was trained on low distributed ***B***-class data performs much better. Combining the highly distributed ***C***-class data with lower distributed data (***A***, ***B***, ***D***) could give the trained networks (***6***, *7*) better performance on ***C***-class prediction, but it worsened the prediction for other classes. It is clearly seen that the data distribution significantly affected the network's performance. Therefore, the multiple networks in implementation were the combination of networks that gave the least value of RMSE in each class, as indicated by the green boxes in Table 3. Based on these comparison results, the multiple networks then consisted of network ***1*** for ***A***-class prediction, network ***2*** for ***B***-class prediction, network *7* for ***C***-class prediction, and network ***4*** for ***D***-class prediction.

To implement a multiple-network scheme, the switching law was needed to select the network when receiving the predicted class from the classifier. For example, if the classifier predicts the ***B***-class, the switching law will select network ***2*** as the launch

**Table 3** Comparisons of the predictive performance between all alternative networks in term of the RMSE values

| Networks | Datasets | RMSE (meters) | RMSE (meters) | | | |
|---|---|---|---|---|---|---|
| | | | ***A*** | ***B*** | ***C*** | ***D*** |
| *1* | *A* | 42.5 | 42.5 | - | - | - |
| *2* | *B* | 13.5 | - | 13.5 | - | - |
| *3* | *C* | 111.6 | - | - | 111.6 | - |
| *4* | *D* | 74.4 | - | - | - | 74.4 |
| *5* | *A+B* | 77.7 | 82.8 | 72.3 | - | - |
| *6* | *C+D* | 101.1 | - | - | 96.0 | 106.0 |
| *7* | *A+B+C+D* | **80.4** | 81.9 | 77.0 | 80.3 | 82.1 |
| Multiple Networks | | **59.1** | 42.5 | 13.5 | 80.3 | 74.4 |

point predictor model, or if the classifier predicts the ***C***-class, the switching law will select network *7*. The overall prediction results across the testing dataset could be obtained by combining the results of all selected networks. As shown in the last row in Table 3, the RMSE value of the multiple networks is 59.1 m, considerably reduced by 26% compared to the single network *7*.

In the case where the classifier accuracy is not perfect, misclassification can deteriorate the overall predictive performance. For example, suppose the classifier misclassifies the actual B-class as the A-class, the switching law will select network ***1*** instead of the expected network **2**, which leads to a worse prediction. Additionally, the accuracy of the launch point prediction can also be affected by the performance of the filter. Poor filtering will output a poorly estimated trajectory which deviates more from the actual trajectory and consequently gets worse in launch point prediction. The effect of the filter and the classifier on the results of launch point prediction were omitted here. However, it is said that all three parts of the proposed method - the filter, classifier, and launch point predictor - are critical to the accuracy of launch point prediction.

Figure 4 compares the launch point prediction of network ***7*** with the individual networks (***1***, ***2***, ***3***, ***4***). It can be seen that all the individual networks, except for network ***3***, make more accurate predictions for the launch points compared to the predictions from network *7*. Notably, network ***2*** stands out from all other networks in terms of the dispersion of predicted launch points. Network ***2*** predicts most of the launch points close to the actual launch points, while the other networks predict more launch points located along the path between the true launch points.

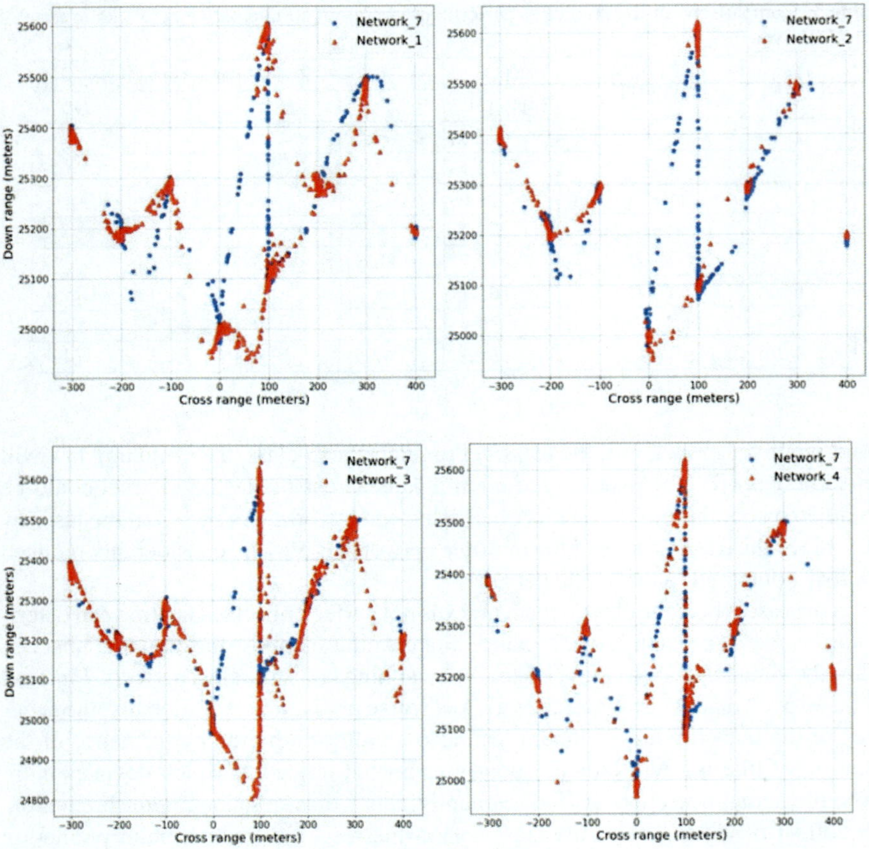


**Fig. 4** Comparison of launch point predictions between network *7* and the individual networks (*1, 2, 3, 4*). The prediction by network *7* is represented as blue dots. The predictions by networks *1, 2, 3,* and *4* are represented as red dots. The coordinates (x, y) of nine actual launch points are (25,000,0), (25,100,100), (25,200,400), (25,200,-200), (25,300,200), (25,300,-100), (25,400,-300), (25,500,300), and (25,600,100)

## 4 Conclusion

In this paper, the method for improving the accuracy of launch point prediction for projectile targets using multiple LSTM networks was presented. The method consisted of three parts. The first part involved using the LSTM-based filtering model to filter the noisy radar data and estimate the flight trajectory. The second part involved using the LSTM-based classifier model to identify the class of munitions based on its flight trajectory. Finally, multiple LSTM networks were developed as the launch point predictor model. The switching law was established based on the classification result, which enabled the selection of the most suitable launch point predictor network for each class of munitions. The results showed a considerable improvement in the accuracy of launch point prediction when using multiple LSTM networks compared to a single LSTM network. Furthermore, the predictive performance was found to be inherently correlated to the distribution of the data.

**Acknowledgement** This original work forms part of a CS/AH2 light howitzer project of Defence Technology Institute, the support and encouragement of which is herein acknowledged.

## References

1. Weapons Locating & Counter Battery Radars. https://www.defenseadvancement.com/suppliers/counter-battery-radars/. Accessed 28 Dec 2022
2. Eric, N., Meir, P., Stantio, M.: Projectile launch point estimation from radar measurements. In: 2005 American Control Conference, vol. 2. pp. 1275–1282 (2005)
3. Farina, A., Timmoneri, L., Vigilante, D.: Classification and launch-impact point prediction of ballistic target via multiple model maximum likelihood estimator (MM-MLE). In: 2006 IEEE Conference on Radar, pp. 802–806 (2006)
4. Ravindra VC, Bar-Shalom Y, Willett P (2010) Projectile identification and impact point prediction. IEEE Trans Aerosp Electron Syst 46(4):2004–2021
5. Carpenter, M., Speakman, N., Hartfield, R.J.: Rapid characterization of munitions using neural networks. In: AIAA Atmospheric Flight Mechanics Conference, American Institute of Aeronautics and Astronautics (2016)
6. Carpenter, M., Hartfield, R.J., Zhou, L., Speakman, N.: Statistical learning for munition trajectory prediction. In: AIAA Scitech 2019 Forum, American Institute of Aeronautics and Astronautics (2019)
7. Eckert, J., Carpenter, M., Hartfield, R., Cervantes, N.: Classification of intermediate range missiles during launch. In: AIAA Scitech 2020 Forum, American Institute of Aeronautics and Astronautics (2020)
8. Kim, J., Lim, M.C., Park, S.-S., Kim, I., Choi, H.-L.: Ballistic object trajectory and launch point estimation from radar measurements using long-short term memory networks. In: 2019 7th International Conference on Robot Intelligence Technology and Applications (RiTA), pp. 26–31 (2019)
9. Hou, L., Liu, H.: An end-to-end LSTM-MDN network for projectile trajectory prediction. In: Cui, Z., Pan, J., Zhang, S., Xiao, L., Yang, J. (eds.) Intelligence Science and Big Data Engineering. Big Data and Machine Learning. IScIDE 2019. Lecture Notes in Computer Science, vol. 11936, pp. 114–125. Springer, Cham (2019). https://doi.org/10.1007/978-3-030-36204-1_9
10. Jacobs RA, Jordan MI, Nowlan SJ, Hinton GE (1991) Adaptive mixtures of local experts. Neural Comput 3(1):79–87
11. Nguyen HH, Chan CW (2004) Multiple neural networks for a long term time series forecast. Neural Comput Appl 13(1):90–98
12. Jiang, C., et al.: A MEMS IMU De-noising method using long short term memory recurrent neural networks (LSTM-RNN). Sensors **18**(10), 3470 (2018)
13. Charubhun, W., Chusilp, P.: Development of automatic firing angle calculation for ground to ground MLRS. In: 2015 Asian Conference on Defence Technology (ACDT), pp. 17–26 (2015)
14. Guyon, I., Laboratories, T.B.: A scaling law for the validation-set training-set size ratio (1997)

# Accommodation Descriptions that Influence Airbnb Occupancy Rate Using Ontology

**Rattapon Choogortoud, Dittapol Muntham, Worawek Chuethong, Sart Srisoontorn, Orasa limpaporn, Nathaporn Utakrit, Kanchana Viriyapant, and Nalinpat Bhumpenpein**

**Abstract** The purpose of this study is to determine the occupancy rate influential model on the Airbnb platform by mapping accommodation descriptions to ontology. This research presented the model and developed an ontology of Bangkok travel information. It defined topics at the start of ontology processing to find the topic that influence on the occupancy rate the most. In this research, the outstanding characteristics of the studied city were identified, defined and categorized by creating the Bangkok Travel Information Ontology to determine the average occupancy rate that occurs when writing each accommodation description to prove whether the strengths are presented in writing or not and to be a tool to help renters on the online platform rent rooms. It can be applied to set a strategy for writing descriptions of various strengths according to ontology and presenting data analysis results through data visualization. The hidden topics described in the Bangkok Travel Ontology are important and affect the rate of room rental on Airbnb, which varies in percentage

R. Choogortoud · D. Muntham · W. Chuethong · S. Srisoontorn · O. limpaporn · N. Utakrit · K. Viriyapant · N. Bhumpenpein (✉)
Faculty of Information Technology and Digital Innovation, King Mongkut's University of Technology North, Bangkok, Thailand
e-mail: nalinpat.b@itd.kmutnb.ac.th

R. Choogortoud
e-mail: rattapon.c@email.kmutnb.ac.th

D. Muntham
e-mail: dittapol.m@email.kmutnb.ac.th

W. Chuethong
e-mail: worawek.c@email.kmutnb.ac.th

S. Srisoontorn
e-mail: sart.s@email.kmutnb.ac.th

O. limpaporn
e-mail: orasa.l@email.kmutnb.ac.th

N. Utakrit
e-mail: nathaporn.u@itd.kmutnb.ac.th

K. Viriyapant
e-mail: kanchana.v@itd.kmutnb.ac.th

 
P. Meesad et al. (eds.), *Proceedings of the 19th International Conference on Computing and Information Technology (IC2IT 2023)*, Lecture Notes in Networks and Systems 679,
https://doi.org/10.1007/978-3-031-30474-3_11

depending on the style of writing accommodation descriptions. In addition, there is a difference in average occupancy considering room type in different subjects and formats, with room type "entire home/apt" having a maximum booking average of 40% occupancy in ontology level 1 and 100% in ontology level 2. For room types, the hotel room type is a maximum booking average rental rate of 83.33% in ontology level 1 and 100% in the ontology level 2.

**Keywords** Machine Learning · NLP · Ontology · Data Analytics · Airbnb

## 1 Introduction

A number of studies have been conducted on Airbnb data to generate decision-making for hosts and customers. In 2021, the Airbnb platform, which is part of the P2P business, had more than 4 million hosts in more than 220 countries [1]. Many studies have reported that the impact of a listing description was considered using a sentiment score with NLP based on the AFINN dictionary and text length. Besides, other factors such as reviews and the number of amenities were determined by the influence of a host's occupancy rate [2]. Previous research on host descriptions revealed that constraining descriptions (such as house rules) were used to regulate guest behaviors, which decreased guest satisfaction [3]. Many studies indicated that reviews of customers by machine learning techniques, such as consumer rental decisions [4] or cost predictions [5] with various techniques, were beneficial to help hosts, customers and Airbnb platforms. Most studies considered cost and customer satisfaction [6–8]. However, the host descriptions were only little research to guide host strategies. Thus, the host descriptions with strategies benefit to hosts and customers; the customers will know the prominent point that they appropriated from host descriptions while the host descriptions will show the advantage attributes.

## 2 Literature Review

The use of ontologies in tourism has garnered increasing interest in recent years. Room owners, or lessors, can rent out their properties to tourists, which benefit for the tourism industry in the city where the property is located. Different cities and countries have unique selling points, such as specific features of accommodations and high guest satisfaction rates. Bangkok and its vicinity are tourist destinations in Thailand, with Bangkok being the most visited city in the world (total international travelers: 22.78 million) [9]. Bangkok and its vicinity, which have strengths in culture, food, and services, are popular tourist destinations. To capitalize on these strengths and attract renters, a description of the room highlighting the city's unique features in their descriptions, and it can be shown and proved that the highlights generated by the ontology method affect the rental. It will be able to create a format for writing data. Previously, the ontology was used to populate a Thai tourism ontology from texts using machine learning [10]. Moreover, several studies are leading to ontologies being applied to create recommendation systems [11]. Furthermore, a previous study that utilized Natural Language Processing (NLP) Latent Dirichlet Allocation found that hidden topic analysis affected occupancy rates. If ontology enhances the tourism experience in Bangkok, it will become possible to apply this research to other places. By using data to create an ontology in the same format, lessors can plan to write their host descriptions more effectively. The research of ontologies in the context of room rentals is a promising area of research that can potentially enhance the tourism industry. Nonetheless, further research is warranted to explore the full potential of ontology in tourism and room rentals, as well as to evaluate the strengths and weaknesses of different methodologies.

## 3 Method

The methodology is presented by "The Occupancy Rate Influential Detections Framework" (see Fig. 1) is a comprehensive framework consisting of five blocks and one step for visualization of well-formed data to achieve its intended purpose. The primary objective of the first block referred to as "Block A," is to facilitate the collection of raw data. The framework's specific detection language is thoroughly explained in Sects. 3.1 and 3.2. The subsequent block, labeled "Block B," focuses on text processing and cleansing performed on Data Set B. This step is instrumental in facilitating the subsequent phases of the framework. It ensures that the data obtained is of optimal quality and can be efficiently utilized for the ontology creation process in Block D. Block C is aimed at processing calendar data from Airbnb to obtain the year, month, and availability status of each room in a month. This is performed on Data Set D. Block D is aimed at building ontologies, which are elaborated upon in Sect. 3.3. This block constitutes a vital aspect of the framework as it facilitates the creation of a comprehensive ontology that can be utilized to enhance its effectiveness. The ontology creation process in Block D is executed by utilizing the data obtained and processed in the B blocks.

Block E consolidates the data from Blocks C and D, which is performed on Data Set E. This data set is used for the analysis process, and the output data is well-formed to present the visualization, which is elaborated upon in Sect. 3.4.

### *3.1 Data Collection*

The data was collected from Airbnb [12], which is open data. The Bangkok data set includes 17,074 rooms, hosting descriptions with English 13,295 rooms, whereas 2,874 rooms have multiple languages and non-hosting descriptions in 905 rooms. The description explained the room's quality. Calendar data consisted of 6,231,645

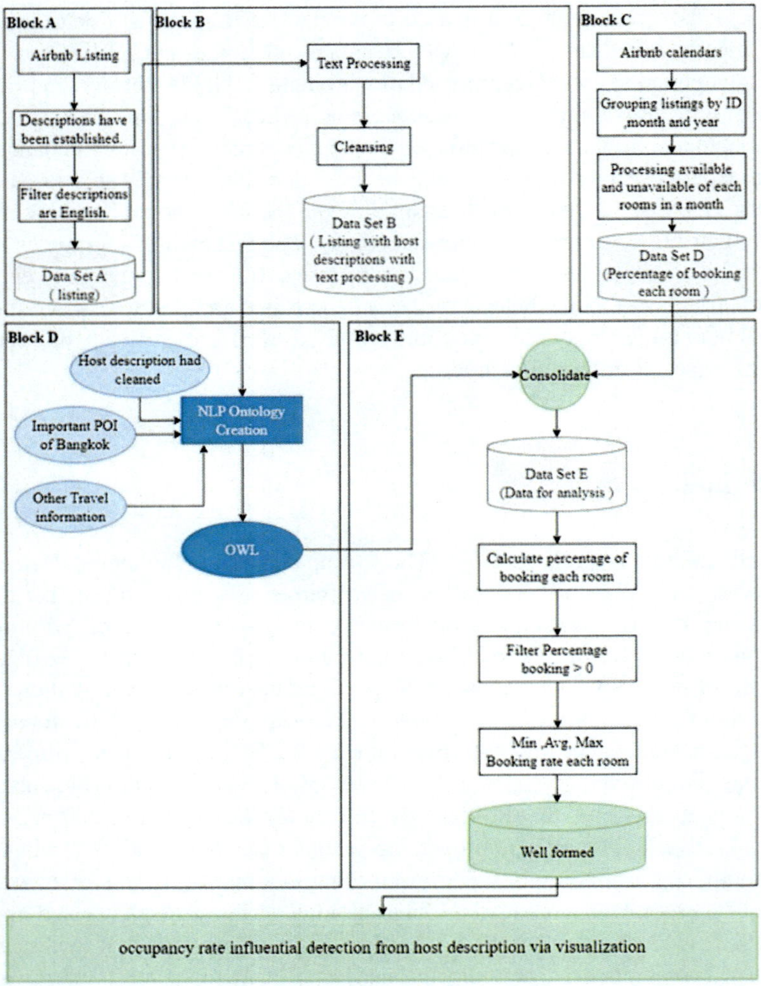


**Fig. 1** Occupancy Rate Influential Detections Framework

lists that were conducted to find the occupancy rate of the rented room on Bangkok's Airbnb platform from 20 June 2022 to 20 June 2023.

## *3.2 Language Detection*

The accommodation descriptions were written in many languages. Thus, after filtering the description, the detected language technique was used in Block A (see Fig. 1) to detect accommodation descriptions with ISO 639–1 language codes and

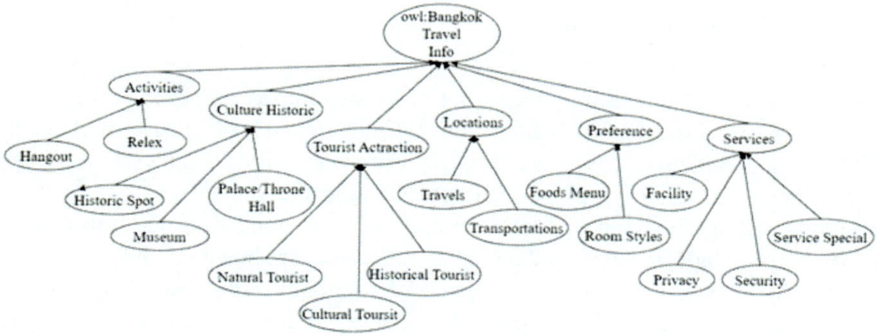


**Fig. 2** Bangkok Travel Information Ontology

ISO 639–1 language names. The accommodation descriptions in only English were taken to Data Set A.

## 3.3 NLP Ontology Creation

In Block B (see Fig. 1), text processing consisted of cleansing by removing noise, tidying text, and recording the accommodation description in Data Set B (see Fig. 2).

Bangkok Travel Information Ontology was created from Important POI of Bangkok with travel information that was conducted to process with accommodation descriptions after cleansing, see Block D (see Fig. 1). After that, the OWL topics consisting of 6 groups, 16 tags and 714 keywords were processed by Ontology. As a result, the possibility of accommodation descriptions was 6 groups and 16 tags.

## 3.4 Consolidate Data to Well-Formed

Consolidating data to well-formed was a process in Block C (see Fig. 1) that processed calendar data and recorded in data set D. Then, the data set C and set D were consolidated with OWL and recorded in data set E. The percentage of bookings for each room was calculated and filtered to be greater than zero. In addition, the minimum, average and maximum booking rates were calculated and stored in data set E. The next process name is "Visualization of Occupancy Rate Influential Detection". This step used a well-formed data set to visualize and interpret influential detections from ontology.

## 4 Result and Discussion

The results showed that different levels of ontology effected on occupancy rates in each room type from 17,074 listing rooms. There were 13,295 interesting rooms with accommodation descriptions in English. From our overview analysis, we found that the booking average of rooms with an accommodation description was 12.40%, while the average booking of rooms without an accommodation description was 3.93%. In this study, the average booking percentages are shown in Table 1.

The chart below (see Fig. 3) presents a histogram graph that compares the percentage of bookings for different ontology group patterns, separated by room type. The chart illustrates the differences in booking percentages across various room types and the unique hidden topics associated with each ontology group pattern. Further information about each room type is explained below.

1. The result of room type, the maximum booking average of the entire home/apt was 40%. The hidden topics were written that found "Activities", "Culture Historic", "Preference", "Services", and "Tourist Attraction" together in the accommodation descriptions out of the 64 patterns that appeared in the topic writing format divided by groups. The entire home/apt type was a higher average occupation

**Table 1** Comparison of Booking Average Percentages by Room Type

| Room type | Average of percent book |
|---|---|
| Entire home/apt | 14.44 |
| Shared room | 10.65 |
| Private room | 9.00 |
| Hotel room | 8.87 |

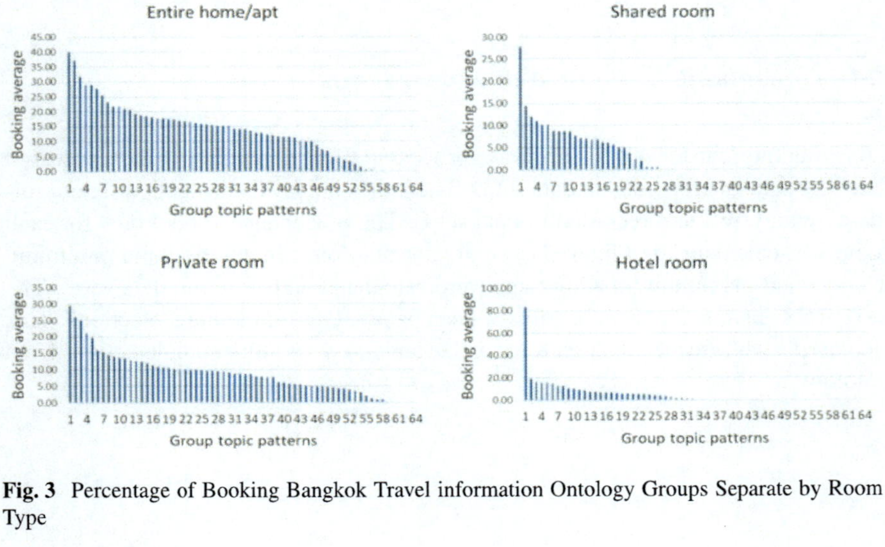


**Fig. 3** Percentage of Booking Bangkok Travel information Ontology Groups Separate by Room Type

rate than the average occupation rate in the group of room types at 14.44% with 30 patterns.

2. The result of room type, the maximum booking average of the shared room was 27.76%. The hidden topics were written that found only "Locations" in the accommodation descriptions. Out of the 64 patterns that appeared in the topic writing format divided by groups. The shared room type was a higher average occupation rate than the average occupation rate in the group of room types at 10.65% with 3 patterns.
3. The result of room type, the maximum booking average of the private room was 29.33%. The hidden topics were written that found "Activities", "Culture Historic", "Preference", "Services" and "Tourist Attraction" together in the accommodation descriptions out of the 64 patterns that appeared in the topic writing format, divided intogroups. The private room type was a higher average occupation rate than the average occupation rate in the group of room types at 9.00% with 30 patterns.
4. The result of room type, the maximum booking average of the hotel room was 83.33%. The hidden topics were written that found "Locations" and "Tourist Attraction" together in the accommodation descriptions out of the 64 patterns that appeared in the topic writing format, divided into groups. The hotel room type was a higher average occupation rate than the average occupation rate in the group of room types at 8.87% with 11 patterns.

The chart above (see Fig. 4) presents a histogram graph that compares the percentage of bookings for different ontology tag patterns, separated by room type. The chart illustrates the differences in booking percentages across various room types and the unique hidden topics associated with each ontology tag pattern. Further information about each room type is explained below.

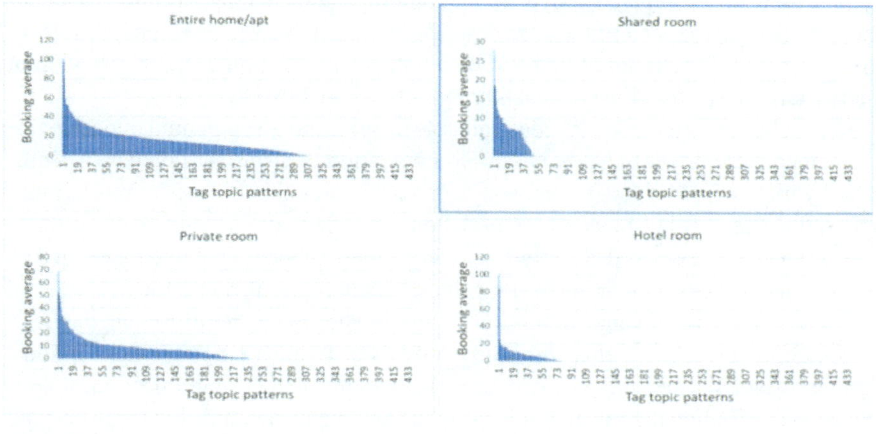


**Fig. 4** The level 2 of Ontology Analysis Showing Sub-topics

1. The room type result for the entire home or apartment type was a maximum booking average of 100%. The tags were written in the hidden topics that showed "Cultural Tourist," "Facility," "Hangout," and "Transportations" together in the accommodation descriptions. Four hundred and fifty patterns were classified by tags that were sub-topics with a higher average occupancy rate than the entire home/apt type at 14.44% with 146 patterns.
2. The room type result for the shared room type was a maximum booking average of 27.76%. The tags were written in the hidden topics that showed only "Transportations" in the accommodation descriptions. The 450 patterns were classified by tags that were sub-topics with a higher average occupancy rate than the shared room type at 10.65% with 9 patterns.
3. The room type result for the private room type was a maximum booking average of 68.33.%. The tags were written in the hidden topics that showed "Cultural Tourist", "Service Special" and "Transportations" together in the accommodation descriptions. The 450 patterns were classified by tags that were sub-topics with a higher average occupancy rate than the private room type at 9.00% with 94 patterns.
4. The room type result for the hotel room type was a maximum booking average of 100%. The tags were written in the hidden topics that showed "Facility", "Palace/Throne Hall/Royal Residence", "Transportations" and "Travel" together in the accommodation descriptions. A total of 450 patterns were classified by tags that were sub-topics with a higher average occupancy rate than the private room type at 8.87% with 30 patterns.

## 5 Conclusion

In this study, the presented framework and ontology detected higher influence occupancy rates from the accommodation descriptions than the average occupancy rate. In addition, the topic patterns could be written for accommodation descriptions without interpretation. The top 5 tags in each type is shown in Table 2.

In conclusion, this study provides valuable insights into the potential of ontologies to enhance the tourism industry pass accommodation by using Airbnb data, specifically by improving the way room owners or lessors, write their descriptions to highlight unique features of their accommodations and the city. The study demonstrates that using ontology-based descriptions can influence booking percentages across different room types, indicating the usefulness of ontologies in this context. However, the study also reveals some limitations, such as the use of multiple languages in descriptions and the fact that this study only used English language descriptions. Further research is needed to fully explore the benefits and limitations of ontologies in tourism, including the use of ontology in different countries and cities with varying languages and contexts.

**Table 2** Top 5 Tags in Each Type

| Top | Entire home/apt |
|---|---|
| 1 | ["Cultural Tourist ","Facility","Hangout","Transportations"] |
| 2 | ["Facility","Hangout","Museum","Service Special"] |
| 3 | ["Cultural Tourist ","Privacy","Relex","Transportations","Travels"] |
| 4 | ["Facility","Museum","Privacy","Room Styles","Transportations","Travels"] |
| 5 | ["Cultural Tourist ","Museum","Transportations"] |
| **Top** | **Shared room** |
| 1 | ["Transportations"] |
| 2 | ["Facility","Privacy","Service Special"] |
| 3 | ["Room Styles","Service Special"] |
| 4 | ["Cultural Tourist ","Facility","Historic Spot","Service Special"] |
| 5 | ["Facility","Travel"] |
| **Top** | **Private room** |
| 1 | ["Cultural Tourist ","Service Special","Transportations"] |
| 2 | ["Security","Transportations"] |
| 3 | ["Relex","Service Special","Transportations","Travels"] |
| 4 | ["Cultural Tourist ","Facility","Historic Spot","Museum","Palace/Throne Hall/Royal Residence"] |
| 5 | ["Facility","Foods Menu","Museum","Room Styles","Service Special"] |
| **Top** | **Hotel room** |
| 1 | ["Facility","Palace/Throne Hall/Royal Residence","Transportations","Travel"] |
| 2 | ["Cultural Tourist ","Transportations"] |
| 3 | ["Facility","Museum","Service Special","Transportations"] |
| 4 | ["Cultural Tourist ","Privacy","Room Styles","Service Special"] |
| 5 | ["Facility","Service Special"] |

## References


1. Georgios, Z., Davide, P., John, W.B.: A first look at online reputation on airbnb, where every stay is above average. Market. Lett. **32**, 1–16 (2020). https://doi.org/10.1007/s11002-020-09546-4
2. Richard, D.M., Anthony, C., Tiffany, K., Lena, T., Nour, A.Z.A.M.: The Impact of an AirBnb host's listing description sentiment and length on occupancy rates. arXiv:1711.09196, 1–17, November (2017). https://doi.org/10.48550/arXiv.1711.09196
3. Lanfei, G., Hui, L., Sai, L., Jingjing, Y., Rob, L.: How does constraining description affect guest booking decisions and satisfaction?. Tourism Manage. **93**, 104607 (2022). https://doi.org/10.1016/j.tourman.2022.104607
4. Muhammad, R.R., Walayat, H., Asaf. V.: Performance analysis of deep approaches on airbnb sentiment reviews. In: the 10th International Symposium on Digital Forensics and Security (ISDFS) (2022)
5. Md DI, Bin L, Kazi SI, Rakibul A, Md RM, Md EH (2022) Airbnb rental price modeling based on latent Dirichlet Allocation and MESF-XGBoost composite model. Mach. Learn. Appl. 7:1–9. https://doi.org/10.1016/j.mlwa.2021.100208

6. Yi L, Rebecca T (2019) Understanding hidden dimensions in textual reviews on Airbnb: an application of modified latent aspect rating analysis (LARA). Int. J. Hosp. Manage. 80:144–154. https://doi.org/10.1016/j.ijhm.2019.02.008
7. Mohamed, C., Omar, B., Younes, C.: Towards a machine learning and data mining approach to identify customer satisfaction factors on Airbnb. In: the 7th International Conference on Optimization and Applications (ICOA) (2021). https://doi.org/10.1109/ICOA51614.2021.9442657
8. Manojit C, Subrata KM (2019) Do airbnb host listing attributes influence room pricing homogenously? Int. J. Hosp. Manage. 81:54–64. https://doi.org/10.1016/j.ijhm.2019.03.008
9. The world's most visited cities. https://travelness.com/most-visited-cities-in-the-world. Accessed 09 Nov 2022
10. Imsombut, A., Sirikayon, C.: An alternative technique for populating Thai tourism ontology from texts based on machine learning. In: Proceedings of 15th International Conference on Computer and Information Science (ICIS), Okayama, pp. 1–4 (2016)
11. Lee, C.I., Hsia, T.C., Hsu, H.C., Lin, J.Y.: Ontology-based tourism recommendation system. In: 4th International Conference on Industrial Engineering and Applications (ICIEA), 21–23 April 2017
12. Inside Airbnb. http://insideairbnb.com/get-the-data/. Accessed 02 Dec 2022

# Comparison of Data Augmentation Techniques for Thai Text Sentiment Analysis

**Kanda Rongsawad and Watchara Chatwiriya**

**Abstract** Many text corpora e.g., Thai text, have problems with low amounts of data or imbalanced data. Data augmentation is one technique used to solve this problem by generating synthetic data. Various data augmentation techniques are applied to text, each method has different algorithms to generate new text. This paper will compare the performance of three well-known data augmentation techniques: Easy Data Augmentation, Embedding Replacement, and Replacement by the masked language model. We apply these data augmentation techniques to the Thai text sentiment corpus to generate synthetic text, then the synthetic text will be used to train the sentiment classifier. After that, the performances of classifiers are analyzed and compared. Three sentiment classifiers used in this paper are Convolutional Neural Network (CNN), Bidirectional Long-Short Term Memory (Bi-LSTM), and Gated Recurrent Unit (GRU). From the results, we found that data augmentation techniques can improve sentiment classifiers' performance. The EDA technique outperformed the others in the Cosmetics Drink and All topics sets but in Restaurant sets, MLM is the best technique that can improve Thai sentiment classification performance. The results show that EDA can be used as a baseline for further study of data augmentation.

**Keywords** Text Sentiment analysis · Easy Data Augmentation · Embedding Replacement · Replacement by the masked language model

## 1 Introduction

Most text corpora, especially low-resource language corpora, have problems with low amounts of data or imbalanced data. As a result, the amount of data is insufficient

K. Rongsawad (✉) · W. Chatwiriya
School of Engineering, King Mongkut's Institute of Technology Ladkrabang, Bangkok, Thailand
e-mail: 59601307@kmitl.ac.th

W. Chatwiriya
e-mail: watchara.ch@kmitl.ac.th



P. Meesad et al. (eds.), *Proceedings of the 19th International Conference on Computing and Information Technology (IC2IT 2023)*, Lecture Notes in Networks and Systems 679,
https://doi.org/10.1007/978-3-031-30474-3_12

to train large and complex models like deep learning models. Thai text corpora also have a problem with low amounts of data as well. Additionally, the text in corpora may contain misspellings or variations in writing styles or language registers. This may cause the text to have complex meanings or can be interpreted in many meanings resulting in ambiguity. Generating synthetic text by modifying some parts of text or synthesizing new text is one approach that can be used to solve problems with low amounts of data. This technique is called Data Augmentation [1]. There are many data augmentation techniques have been applied to English texts that have yielded satisfactory results such as Synonym Replacement, Embedding Replacement, and Replacement by masked language model [2].

This paper will apply these data augmentation techniques to the Thai text corpus used for sentiment analysis. Then, the performances of data augmentation were compared with the classification results of the models trained with synthetic text generated from different data augmentation techniques to determine which technique is most suitable for the Thai text corpus.

## 2 Literature Review

### *2.1 Problems of Text Corpora*

The text corpora often have problems with First, the number of texts is low and insufficient for training the models, especially the low-resource language corpora. Second, data are imbalanced because each class has different occurrences. Some classes may have a low probability of occurrence, resulting in lower amounts of data than other classes. Third, texts in corpora have low-quality, misspelled words and are written in a variety of writing styles or language registers. This problem is often found in texts collected from the Internet or social networks. And fourth, the variety of vocabulary in the corpus does not cover all the vocabulary in the language.

One approach that can be used to solve the first and second problems is to generate synthetic text with data augmentation techniques. There are well-known data augmentation techniques applied to English text. For example, the synonym replacement method [2] generates a new text by replacing words in the text with its synonym. Easy Data Augmentation (EDA) [3], introduced by Wei, J. and Zou, K. in 2019, is one method that uses synonym replacement to generate new text. In addition, EDA also adds three other operations to generate text: 1 Random Insertion 2 Random Swap 3 Random Deletion. The Embedding Replacement [2] generates a new text with word replacement that is like synonym replacement. But this method uses the pre-train word embedding model to find a new word that is the nearest word or most similar instead of the synonym used in the synonym replacement. And Replacement by masked language model [2] generates a new text with new words replacement as the previous methods. Both previous methods find new words by considering only the word that will be replaced, but this method finds new words based on surrounding words instead. We'll replace the word we want to replace with the <mask> and use the masked language model to predict which words should be filled in the <mask> position.

### 2.2 *Text Sentiment Analysis*

Text sentiment analysis is the analysis of the content of texts to interpret the feelings, emotions, or intentions that the author wants to express. This information can provide an overview of how the people's attitudes toward the topic being written about. For example, a customer's comment about a restaurant can be used to analyze how this customer feels about this restaurant. And when we have the comments of many customers, we can analyze them to find an overview of the customers' opinions about this restaurant. In addition, if we analyze further, we may be able to know which services are popular or need to be improved. These analysis results can be used in business planning to increase the value of a business. For this reason, text sentiment analysis is one problem researchers are interested in.

In the Thai language, there are studies on Thai text sentiment analysis as well, for example, in 2022, Pasupa, K. and Seneewong Na Ayutthaya, T. [4] analyzed sentiment from texts of three corpora namely 1) Thai children's tales, 2) Thai Economy Twitter, and 3) Wisesight. The results show that the Bi-LSTM-CNN model, a hybrid model between the Bi-LSTM and CNN, can analyze sentiment most efficiently. The F1-score were 0.74, 0.77, and 0.55 respectively. And in the same year, Srikamdee, S., Suksawatchon, U. and Suksawatchon, J. [5] used texts from two corpora: 1) Wisesight and 2) Thai children's tales to analyze sentiment. The Logistic Regression model and SVM model have the best accuracy.

## 3 Research Methods

The methods used in this research are divided into 4 main parts: 1) Thai text corpus preprocessing. This step will verify, correct misspelled words, and discard ambiguous text. 2) Thai text tokenization. This part divides into 2 types: text tokenization for generating synthetic texts and text tokenization for sentiment analysis. 3) Synthetic texts Generation by data augmentation. In this step, synthetic texts are generated with different data augmentation techniques. These synthetic texts will be used to train the sentiment classifier in the next parts. The last part is 4) Text sentiment classification. The details are as follows.

### 3.1 Thai Text Corpus Preprocessing

The research used a sentiment analysis corpus named Wisesight [6]. The data in this corpus are Thai texts collected from social media from 2016 to 2019. There are 26,737 messages and 4 categories: positive, negative, neutral, and questions. Most messages are opinions about products or services of 1) restaurants, 2) cosmetics, 3) drinks, 4) cars, and 5) hotels. And the remainder is opinions about the situation that occurred at the time the data was collected. There are main text preprocessing steps are as follows: 1) Only positive and negative sentiments were used so that the sentiment classes were the same as [3]. 2) Ambiguous texts were discarded. 3) Texts were grouped according to the topic of comment and selected only the text related to restaurants, drinks, and cosmetics to use. 4) Texts written with casual or semi-formal styles were selected. 5) Misspelled words and transliteration errors have been corrected. 6) Emojis, emoticons, and special characters have been removed. 7) Texts have been adjusted to a maximum of approximately 400 characters.

After the text has been preprocessed, four data sets are used in this paper: Cosmetics set contains 615 messages, Drinks set contains 593 messages, Restaurants set contains 750 messages, and All topics set contains 1,994 messages, obtained by combining Cosmetics, Drinks, and Restaurants.

### 3.2 Thai Text Tokenization

The Thai text tokenization is divided into two parts. The first part is text tokenization for generating synthetic texts and the second part is text tokenization for sentiment analysis. Both parts use the Maximum Matching algorithm (newmm tokenizer), a dictionary-based tokenizer via the PyThaiNLP library [7]. But in the first part, the dictionary named DictA was used for generating synthetic texts. DictA is obtained by combining the vocabulary of 3 dictionaries: 1) PyThaiNLP's Thai words 2) The Royal Institute Dictionary and 3) Thai-English Lexitron Dictionary 2.0 and name entities of government agencies, universities, administrative districts, roads, and bridges from NaNa Wisamanyanam of the Royal Institute to provide the most comprehensive vocabulary. This will help the tokenized words to have a complete meaning. And the second part, the dictionary named DictB was used for sentiment analysis. This dictionary is vocabs in a word embedding model called thai2fit [8]. DictB is used words to ensure that tokenized words can be correctly transformed into word vectors.

### 3.3 Synthetic Texts Generation by Data Augmentation

Three data augmentation techniques: Easy Data Augmentation (EDA), Embedding Replacement (ER), and Replacement by masked language model were used to

**Table 1** Example of texts generated with EDA. The origin text is " เยอะมาก น่าทานทั้งนั้นเลยค่ะ"

| Operation | Example |
|---|---|
| Synonym Replacement | เยอะมาก น่า**รับประทาน**ทั้งนั้นเลยค่ะ |
| Random Insertion | เยอะมาก น่าทาน**มากมาย**ทั้งนั้นเลยค่ะ |
| Random Swap | เยอะมาก น่า**ทั้งนั้น**ทาน**เลย**ค่ะ |
| Random Deletion | เยอะมากน่าทาน[**ทั้งนั้น**]เลยค่ะ |

generate synthetic texts that will be used to train the sentiment classifier. Details are as follows.

**Easy Data Augmentation (EDA)**

EDA was introduced by Wei, J. and Zou, K. in [3]. EDA was applied to English text classification corpora. The results showed that models trained with EDA-generated data had an average accuracy increase of 3.0% and 0.8% for the training set sizes 500 and full datasets respectively. This shows that EDA can be applied to English text and should be able to apply to Thai text as well. There are four operations to generate synthetic texts. 1) Synonym Replacement generates new text by replacing random words with their synonyms in the text. 2) Random Insertion generates new text by finding synonyms of random words in the text and inserting them into random positions in the text. 3) Random Swap generates new text by swapping the position of two random words in the text. 4) Random Deletion generates new text by randomly removing words in the text. In this paper, we want to test the efficiency of EDA when applied to Thai texts, so we will be used only Thai words to find synonyms in Synonym Replacement and Random Insertion operations (Table 1).

**Embedding Replacement (ER)**

The Embedding Replacement (ER) uses the pre-train word embedding model to find new words to replace words in the text. First, transform a selected word into the word vector with word embedding, then this words vector is used to calculate cosine similarity with all vocabulary in the word embedding model, and the word with the most similarity and the same part of speech as an origin word is selected to replace an origin word. After that, the new text will be checked for duplicates. If this text is duplicated, we will discard this text. The word embedding model called thai2fit [8] is used. thai2fit is a pre-train model trained with Thai Wikipedia Dump data.

**Replacement by Masked Language Model (MLM)**

Replacement by masked language model (MLM) uses a masked language model to predict new words to replace the origin words in the text. In Synonym Replacement of EDA and ER, new words are selected based on origin words, but this method finds new words based on surrounding words instead. First, a word in the text is selected randomly, then it is replaced with <mask>. After that, the masked language model is used to predict which words should be filled in the <mask> position. Finally, the new text will be checked for duplicates. If this text is duplicated, we will discard this text.

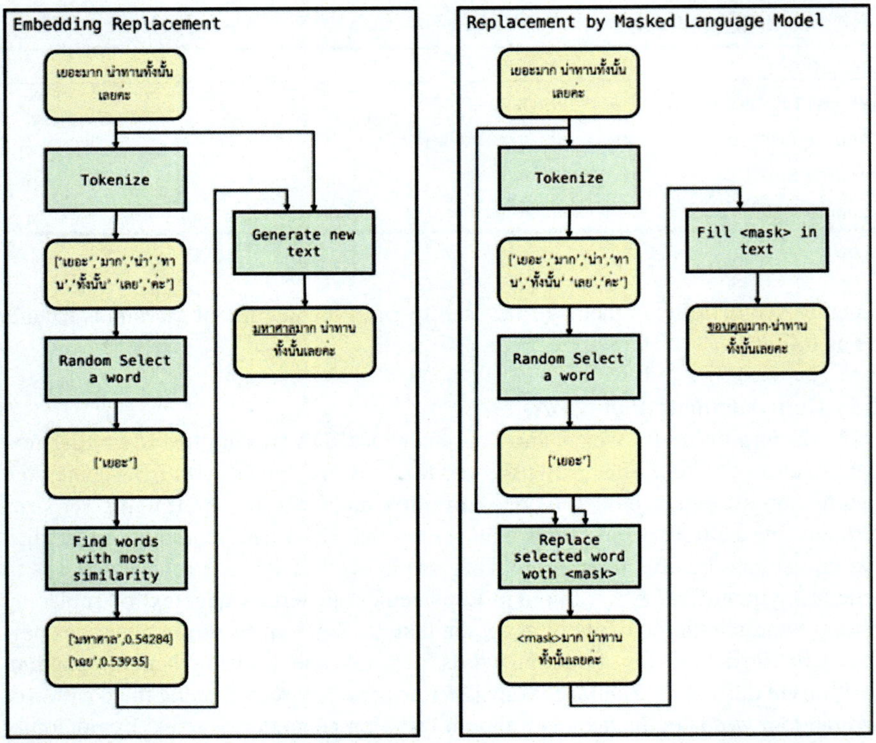


**Fig. 1** Embedding Replacement (left) and Replacement by masked language model (right)

The masked language model called WangchanBERTa [9] is used. WangchanBERTa is a pre-train Thai language model based on RoBERTa [10] (Fig. 1).

## 4 Text Sentiment Classification

Three classifiers used for sentiment classification are Convolutional Neural Networks (CNN), Bidirectional Long-Short Term Memory (Bi-LSTM), and Gated Recurrent Unit (GRU). The CNN and Bi-LSTM model structure was used according to [3], but the word embedding model was changed from GloVe, which does not support the Thai language, to thai2fit [7], which supports the Thai language instead. GRU uses the same structure as Bi-LSTM, but the Bi-LSTM layers are replaced with GRU layers. The structures of CNN, Bi-LSTM, and GRU models are shown in Fig. 2.

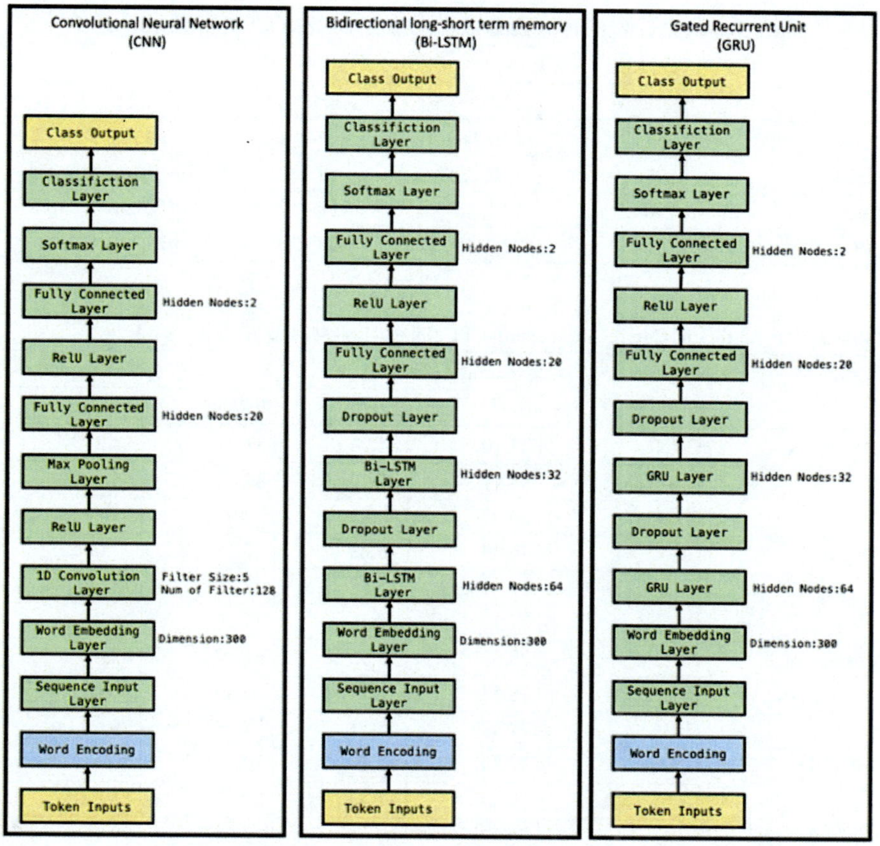


**Fig. 2** Structure of CNN, Bi-LSTM, and GRU models

## 5 Analysis of Experimental Results

We apply EDA, ER, and MLM techniques to 4 data sets (Cosmetics, Drink, Restaurant, All topics) to generate synthetic text. one original text is used to generate 16 synthetic texts that were the same for all 3 techniques. We select only the generated texts that have unique 16 patterns for ER and MLM methods. The numbers of synthetic text generated by EDA, ER, and MLM are presented in Table 2.

After that, both origin and synthetic text were used to classify sentiment from text with CNN Bi-LSTM and GRU models and 5-Fold Cross Validation. The synthetic texts used to train the model must be generated from the original texts in the training set, and the texts used to test the model must be the original texts of the test set.

From the results in Table 3, we found that in the Cosmetics Drink and All topics set, the best accuracy was obtained by models trained with texts generated with EDA. But in the Restaurant set, the best accuracy was obtained by models trained with texts generated with MLM. We assume that the EDA technique can't generate

**Table 2** The number of synthetic texts generated by EDA, ER, and MLM methods

| Topic | Data Size | | | |
|---|---|---|---|---|
| | Origin | EDA | ER | MLM |
| Cosmetics | 651 | 10,416 | 9,712 | 10,224 |
| Drink | 593 | 9,488 | 8,352 | 9,168 |
| Restaurant | 750 | 12,000 | 11,456 | 11,856 |
| All topics | 1994 | 31,904 | 29,520 | 31,248 |

**Table 3** The sentiment classification results of CNN, Bi-LSTM, and GRU

| Topic | Model | Accuracy (%) | | | |
|---|---|---|---|---|---|
| | | Base | EDA | ER | MLM |
| Cosmetics | CNN | 85.10 | 84.94 | 83.71 | 84.33 |
| | Bi-LSTM | 85.41 | **87.10** | 84.03 | 85.10 |
| | GRU | 80.95 | 81.87 | 80.02 | 83.41 |
| Drink | CNN | 76.39 | **78.43** | 77.25 | 77.08 |
| | Bi-LSTM | 75.71 | 77.38 | 76.71 | 75.86 |
| | GRU | 71.17 | 75.56 | 75.56 | 76.73 |
| Restaurant | CNN | 78.53 | 80.93 | 81.47 | **82.40** |
| | Bi-LSTM | 78.53 | 78.27 | 80.80 | 80.40 |
| | GRU | 77.47 | 78.53 | 77.07 | 79.60 |
| All topics | CNN | 79.34 | **82.10** | 79.44 | 80.44 |
| | Bi-LSTM | 79.44 | 81.44 | 80.39 | 81.34 |
| | GRU | 78.03 | 81.04 | 79.04 | 79.29 |

text effectively because only some vocabulary in Restaurant set has synonyms. This is different from MLM techniques that can generate text more efficiently because they use contextual words to find new words.

The CNN model achieved the highest accuracy in Drink Restaurant and All topics sets. In Cosmetics sets, The Bi-LSTM model is the most accurate. By examining the texts in the Cosmetics set, we found that many words in this set are product names and transliterated words, and these words are often separated into multiple words after tokenization. Since the Bi-LSTM operates in two directions, forward direction and backward direction, and backward direction, it can handle this problem better than other models.

## 6 Conclusion

In conclusion, the sentiment classification results show data augmentation techniques can apply to Thai text. The EDA outperformed the others in Cosmetics Drink and All topics sets. The accuracy of the sentiment classifier train with text generated with EDA was 87.10%, 78.43%, and 82.10%, respectively. In Restaurant sets, MLM is the best technique that can improve sentiment classification accuracy. The accuracy was 82.40%. This means that EDA could be used as a baseline to further study data augmentation for Thai texts. In addition, the data used in the research were opinions about cosmetics, drinks, and restaurants only, so the diversity of the vocabulary was limited. In further research, opinions about other products and services, or topics should be added to increase the diversity of the vocabulary or apply other data augmentation techniques to Thai texts for further studies such as back-translation.

## References


1. Li B, Hou Y, Che W (2022) Data augmentation approaches in natural language processing: a survey. AI Open 3:71–90. https://doi.org/10.1016/j.aiopen.2022.03.001
2. Bayer M, Kaufhold M-A, Reuter C (2022) A survey on data augmentation for text classification. Assoc Comput Mach 55(7):1–39. https://doi.org/10.1145/3544558
3. Wei J, Zou K (2019) EDA: easy data augmentation techniques for boosting performance on text classification tasks. In: Proceedings of the 2019 conference on empirical methods in natural language processing and the 9th international joint conference on natural language processing (EMNLP-IJCNLP), Hong Kong, China. Association for Computational Linguistics, pp 6382–6388. https://doi.org/10.18653/v1/D19-1670
4. Pasupa K, Seneewong Na Ayutthaya T (2022) Hybrid deep learning models for Thai sentiment analysis. Cogn Comput 14(1):167–193. https://doi.org/10.1007/s12559-020-09770-0
5. Srikamdee S, Suksawatchon U, Suksawatchon J (2022) Thai sentiment analysis for social media monitoring using machine learning approach. In: 37th international technical conference on circuits/systems, computers and communications (ITC-CSCC), Phuket, Thailand. IEEE, pp 1–4. https://doi.org/10.1109/ITC-CSCC55581.2022.9894882
6. Suriyawongkul A, Chuangsuwanich E, Chormai P, Polpanumas C (2019) PyThaiNLP/wisesight-sentiment: first release. Zenodo. https://doi.org/10.5281/zenodo.3457447
7. Phatthiyaphaibun W, Chaovavanich K, Polpanumas C, Suriyawongkul A, Lowphansirikul L, Chormai P (2016) PyThaiNLP: Thai natural language processing in Python. Zenodo. https://doi.org/10.5281/zenodo.3519354
8. Polpanumas C, Phatthiyaphaibun W (2021) thai2fit: Thai language implementation of ULMFit. Zenodo. https://doi.org/10.5281/zenodo.4429691
9. Lowphansirikul L, Polpanumas C, Jantrakulchai N, Nutanong S (2021) WangchanBERTa: pretraining transformer-based Thai language models. arXiv. https://doi.org/10.48550/ARXIV.2101.09635
10. Liu Y, et al (2019) RoBERTa: a robustly optimized BERT pretraining approach. arXiv. https://doi.org/10.48550/ARXIV.1907.11692

# Jok Mae Jaem Woven Fabric Motif Recognition Using Convolutional Neural Network

**Yosawimon Attawong, Jakramate Bootkrajang, and Watcharee Jumpamule**

**Abstract** Pha-tho Teen Jok (Teen Jok woven fabric) is a special crafts discipline of Tai ethnic groups. It is one of the traditional handwoven fabrics with pattern, color and texture that represents the cultural wealth of Tai people. Jok Mae Jaem is one of Tai Yuan's signature styles found in Mae Jaem District, Chiang Mai Province, Thailand. It can be further categorized into sixteen different motifs. The subtleties pose challenges in recognizing the motifs especially for those who are unfamiliar with the style. Being able to differentiate Jok Mae Jaem motifs with minimal reliance on the experts is found to be beneficial. In this paper, we studied the possibility of employing convolutional neural network (CNN) models to identify sixteen Jok Mae Jaem motifs. In addition, we mixed in fabrics from other four regions of northern Thailand, namely Si Satchanalai, Nan, Lap-lae and Long to make the dataset more realistic. Data augmentation and transfer learning techniques have also been adopted to create variations in the sample and to facilitate the learning under limited data setting. Experimental results based on a manually collected fabric image dataset showed that the EfficientNet attained superior predictive performance as indicated by accuracy, precision and recall while the ResNet followed closely behind.

**Keywords** Pha-tho Teen Jok · Image classification · Convolutional neural network

## 1 Introduction

Pha-tho Teen Jok is one of the traditional handwoven fabrics with distinctive style, colorful patterns and delicate texture. It is arguably one of the most beautiful woven

Y. Attawong · J. Bootkrajang (✉) · W. Jumpamule
Department of Computer Science, Chiang Mai University, Chiang Mai, Thailand
e-mail: jakramate.b@cmu.ac.th

Y. Attawong
e-mail: yosawimon.attawon@cmu.ac.th

W. Jumpamule
e-mail: watcharee.j@cmu.ac.th



P. Meesad et al. (eds.), *Proceedings of the 19th International Conference on Computing and Information Technology (IC2IT 2023)*, Lecture Notes in Networks and Systems 679,
https://doi.org/10.1007/978-3-031-30474-3_13

fabrics that portrays sweet and emotional moods embracing a simple lifestyle of the Thai Lanna people. Pha-tho Teen Jok is referred to cotton or silk textiles that are hand woven according to the conventional methods, with motifs woven in discontinuous supplementary weft cotton or silk so that it forms patterns of high artistic value. Currently, Teen Jok woven fabric has various types of different motifs and patterns, which spread across many regions of Thailand with their distinct names and meanings. Pha-tho Teen Jok or Pha Teen Jok is a special crafts discipline of Tai Yuan, Tai Phuan, Tai Lue, Tai Krang and Phu Tai peoples. Normally, Pha-tho Teen Jok is a decorative weaving technique using discontinuous supplement wefts. The wefts are horizontal threads woven across the width of fabric, through the lengthwise wrap yarn. This technique consists of carefully inserting weft threads of different colors through the wrap threads to make a complex and colorful pattern [1]. The motifs or patterns on Pha-tho Teen Jok in Thailand vary according to the characteristics of each ethnic and region. Jok Mae Jaem Woven Fabric is one type of Pha-tho Teen Jok which has various kinds of motifs that have been produced from generation to generation. The motif embeds the meaning of the ancestors with links to religious beliefs. It is famous for its colors and delicate patterns [1, 2]. When viewed with the naked eyes, the texture and motif of each Teen Jok woven fabric look pretty similar especially for non-expert. To be able to differentiate the motifs with minimal effort, it is necessary to consider classifying the motifs with the help of computer vision-based technology. Although, the studies of fabric motifs classification on local hand-woven fabrics and batik patterns have received much attention recently, for example [3–7] study that focuses on Jok Mae Jaem fabric is still limited.

Recently, the deep learning approach is being widely used in image classification tasks. Convolutional Neural Network (CNN) is a special type of deep learning model which is very often applied to the image processing field [8]. Nowadays, CNN is widely accepted as one of the primary models for image classification and recognition because of its ability to learn high level features in images and produce highly accurate predictions. Another compelling reason is that it automatically extracts features from input with less human intervention [4, 5] thereby facilitating the realization of the recognition system. There are many CNN architectures that have been well trained to detect and classify images in computer vision. Since the accuracy of deep learning models largely depends on the quality, quantity, and contextual meaning of training data which could not be sufficiently fulfilled in many cases including ours where the sample size of training data is quite limited, in this study we resorted to employing data augmentation and transfer learning in classification model training. Transfer learning is based on the principle of developing learning by using information obtained from a related task in a new task. It was used to speed up the process of training models to get better performance faster and pre-trained deep learning networks are already well-trained using other datasets, which could be refined to obtain high accuracy in our case [5].

In this study we studied three of the widely used deep neural networks. Specifically, we studied ResNet50, MobileNetV2 and EfficientNetV2B3 architectures in addressing the problems of classification of Pha-tho Teen Jok Mae Jaem fabric motifs. The performance of the three architectures was then analyzed and compared.

## 2 Related Work

In recent years, there have been studies on image classification using deep learning with impressive predictive performances. This technique has been proven of its capacity in capturing patterns or motifs in images. Research in textile motif recognition has been done by several researchers to put forward new methods that have better performance than those in the previous studies. Study [6] used CNN for classification of motifs of hand-woven fabrics in Loei Province, Thailand. In the experiment, tenfold cross-validation was employed to train and test 4,500 images in the dataset based on two convolutional neural network models: Inception-v3 and MobileNet. The results revealed that MobileNet outperforms Inception-v3 with the accuracy rate of 98.22% and 93.21% respectively. Research by [7] used convolution neural networks to classify batik textile based on its manufacturing method. Image data augmentation was also used to prevent overfitting as well as to provide variations in the training data. The experimental results with fivefold cross validation showed that Densenet169 gives the best results on the original dataset with an accuracy of 79.17%. With a limited amount of data, transfer learning technique was also considered. All models showed an increase in performance when using the modified dataset. VGG13 shows the best performance on the modified dataset with an accuracy of 87.61% while Densenet169 did not show a significant difference in performance. Batik researchers in [9] used the method of based on a modified VGG architecture to classify batik motifs. Experiments using 2,448 batik images from 5 classes of batik motifs showed that the proposed model has successfully achieved an accuracy of 96.30%. From the work in [10] the researchers applied the convolutional neural network to identify six batik patterns. Overall, 994 images from the 6 classes were collected and then divided into training and test data and image augmentation was also done to provide variations in training data as well as to prevent overfitting. Experimental results on the test data showed that CNN produced an excellent performance as indicated by the accuracy of 94% and top-2 accuracy of 99% which was obtained using the DenseNet network architecture. The work in [11] used 4 simple CNN models namely VGG, ResNet, Inception and EfficientNet to detect patterns with the goal to extract the texture or pattern from the jacquard fabrics. The result showed EfficientNet yielded a high accuracy rate reaching 99% approximately.

## 3 Dataset and Methodology

### *3.1 Dataset*

#### 3.1.1 Data Collection

The dataset used in this research are the Pha-tho Teen Jok fabric images. We collected a total of 800 images of motifs from photo taken with mobile phone and captured from

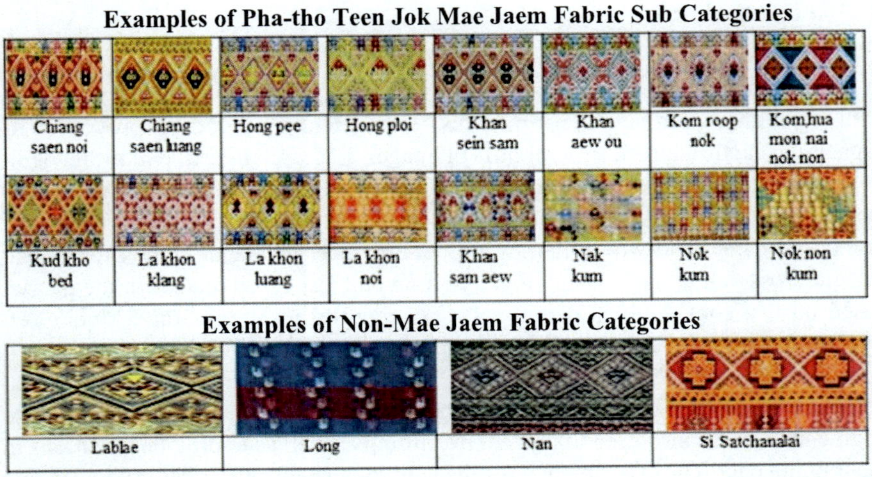


**Fig. 1** Example of Pha-tho Teen Jok fabric motif images: 16 categories of Pha-tho Teen Jok Mae Jaem and the four others

the Internet. The motifs consist of Pha-tho Teen Jok Mae Jaem, Si Satchanalai, Long, Nan and Lap-lae in JPEG format. Each of the collected images has been cropped and resized to $224 \times 224 \times 3$ dimensions. This is done to optimize the training time and the classification accuracy for three models. Unwanted artefacts such as strip, shadow, inconsistent movement, and other disturbances were also treated. In the end, 20 examples of Pha-tho Teen Jok Mae Jaem fabric motif images and other categories of fabric motif images are shown in Fig. 1.

### 3.1.2 Data Preprocessing and Augmentation

At the time of data collection, the dataset consists of 40 examples of each class produce a database of 800 images from 20 different categories of Pha-tho Teen Jok fabric. From the original images, we divided into 16 sub classes of Pha-tho Teen Jok Mae Jaem fabric and the four others. These images are then divided into two subsets namely training and test sets with a ratio of 8:2. Thus, the training and test set consist of 80% and 20% images respectively.

It is well known that deep learning models require a large amount of data for training. In this study, image augmentation is used to increase the amount of data and quality of datasets to evade the problem of overfitting and produce deep learning models with improved accuracy. With image augmentation, during the training process, the image given to the model will be modified with small random transformations without losing its original content. Transformations that can be done include flipping, rotation, zoom, brightness and combinations. A few example images of the augmented Pha-tho Teen Jok Mae Jaem fabric motif dataset are shown in Fig. 2.

**Fig. 2** Examples of image augmentations used in this study

**Table 1** Dataset statistics after data augmentations

| Motif image classes | Training set | Validation set | Test set |
|---|---|---|---|
| Sixteen sub-categories of Mae Jaem fabric | 3,360 | 480 | 960 |
| Si Satchanalai fabric | 210 | 30 | 60 |
| Nan fabric | 210 | 30 | 60 |
| Lap-lae fabric | 210 | 30 | 60 |
| Long fabric | 210 | 30 | 60 |
| **Total** | **4,200** | **600** | **1,200** |

We then augmented the collected images with a total of 4,800 for training images and 1,200 for test images. The training images further divided into a train and a validation set with a ratio of 7:1 (4200:600). Table 1. illustrates the statistics of the split dataset.

## 3.2 *Convolutional Neural Networks Architecture*

Three CNN models were used in the study to classify 20 classes of motifs. These models are the pre-trained ResNet50, MobileNetV2 and EfficientNetV2B3 models. These were chosen because of their philosophy, performance and ability to bring out specific architectural innovations. ResNet50, which was developed at Microsoft and described in 2015, uses a residual model which involves shortcut connections. The pre-trained network is aimed at classifying images into 1000 object categories. As a result, the network has learned rich feature representations for a wide range of images [12]. The architecture of MobileNetV2 is based on an inverted residual structure where the shortcut connections of the residual block are between the thin bottleneck layers. The intermediate expansion layer of the MobileNetV2 uses lightweight depth-wise convolutions in order to filter the features. In traditional residual models, expanded representations in the input are used [13]. It is one of several attempts to reduce the footprint of the computational demanding CNN models. The EfficientNetV2B3 is a new family of convolutional, which use a combination of training-aware neural architecture search and scaling, to jointly optimize training speed and parameter efficiency, the propose is to create networks that have faster training speed and better parameter efficiency than other models [14]. The EfficientNet family is based on a new method for scaling up CNN models. It uses a simple greatly effective compound coefficient and much smaller when compared to other models achieving similar ImageNet accuracy.

### 3.3 Model Training and Evaluation

This experiment is based on the Resnet50, MobileNetV2 and EfficientNetV2B3 models implementation on TensorFlow platform. Even though 3 models were not designed to classify Pha-tho Teen Jok fabric motifs, they should be able to extract useful generic features from Pha-tho Teen Jok fabric images which can be used further for classification. Transfer learning technique can be used to help neural networks learn data representation and transfer knowledge of another pre-trained neural network model to our model [5]. We used models that have been previously trained on the ImageNet dataset to speed up the learning process. We trained the models for 20 learning epochs. The training process was carried out using the same settings in each model to obtain results that can be compared. Adam optimization algorithm was used to optimize the models. Each model is then tested using the test data with a total size of 1,200 images that have been set aside previously. The results of this test will then be summarized in the form of model performance. There are four criteria tested, namely accuracy, precision, recall, and F1-measure.

## 4 Experimental Results and Discussion

Model performance of pre-trained CNN models from ImageNet namely ResNet50, MobileNetV2 and EfficientNetV2 were compared. The same optimization algorithm was applied to each model. In addition, the learning dynamics of these models were evaluated in terms of loss graphs, average training time and model size.

Figure 3, 4 and 5 the trends of training accuracy and validation accuracy, training loss and validation loss. In the graphs, blue and red lines show training accuracy/loss and validation accuracy/loss of the models, respectively. According to Fig. 3, EfficientNetV2B3 has achieved the highest accuracy of 93.88% with a training loss of 0.2119, that is, model learned effectively and it can properly distinguish between sixteen motifs of Pha-Tho Teen Jok Mae Jaem fabric and the four others.

ResNet50 (Fig. 4) has achieved a maximum accuracy of 81.16% with a training loss of 0.6853. The model is effectively less accurate as compared to EfficientNetV2B3. Figure 5 indicates that MobileNetV2 has achieved 42.54% accuracy with a training loss of 1.9503. It shows that the model has learned to some extent but it is the least accurate to classify among sixteen Pha-tho Teen Jok Mae Jaem fabric motifs and the four others. This could be due to the limited complexity of the model which was originally designed for speed and low resource utilization.

Table 2 shows the average training time and model size. The MobileNetV2 network architecture shows the shortest average training time of 67.5 s to complete

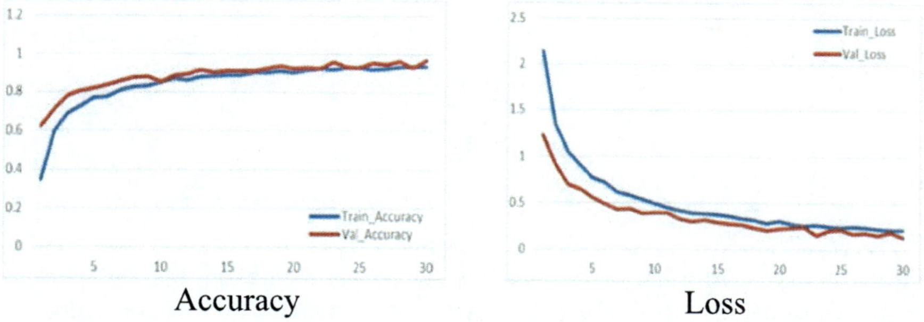


**Fig. 3** Training accuracy and loss for EfficientNetV2B3

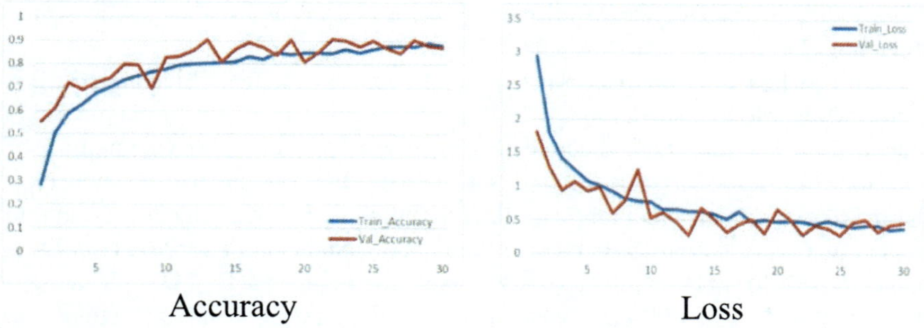


**Fig. 4** Training accuracy and loss for ResNet50

**Fig. 5** Training accuracy and loss for MobileNetV2

on each epoch and model size was the smallest (12.6 MB). ResNet50 model shows the biggest size with the longer average training time when compared with the others.

Table 3 shows the evaluation metrics of both validation set and test set of the three CNN models and the performance of each classification model in term of accuracy, precision, recall, and F1-Score. The experiment is to determine the best model

**Table 2** Average training time size of each model

| Model | Average training time (sec.) | Estimated Total Size (MB) |
|---|---|---|
| MobileNetV2 | 67.5 | 12.6 |
| EfficientNetV2B3 | 146.5 | 50.3 |
| ResNet50 | 307.5 | 92.9 |

with the highest performance metrics. Through experimentation, the best performance was shown by EfficientnetV2B3 with an accuracy of 80.94%, precision of 86.06%, recall of 82.83% and F1-score of 83.01%. Meanwhile the worst performance is shown by MobileNetV2 with accuracy, recall, and F1-measure below 50% and precision above 50%. In addition, ResNet50 model performed fairly well, with above 70% performance across all measurement metrics and higher test loss than EfficientnetV2B3.

Figure 6 shows the confusion matrix of the test set for the EfficientNetV2B3, ResNet50 and MobileNetV2 models. In the confusion matrix, while the diagonal axis number show the number of accurate classifications, the others show the number of inaccurate classifications. The EfficientNetV2B3 was the best model to deal with only Pha-tho Teen Jok Mae Jaem fabric motif recognition and it makes the most accurate prediction. Classification accuracy rates were found to be 98.33% for MJ11: La khon luang, 96.66% for MJ15: Nok kum, MJ16: Nok non kum, 95% for MJ13: Khan sam aew, MJ14: Nak kum, and nearly 90% for MJ1: Chiang saen noi and MJ3: Hong pee, respectively.

To summarize, EfficientNetV2B3 model obtained the highest accuracy, precision, recall, and F1-score values and better performance than the other models on unseen test data. Moreover, the EfficientNetV2B3 was the best model to deal with only Pha-tho Teen Jok Mae Jaem fabric motif recognition. The work in [11] obtained similar results for the EfficientNetV2B3 model. The reason for the superiority of EfficientNetV2B3 architecture could be that it is designed for small dataset. However, the results show that the ResNet50 also have high accuracy, but it provided lower test

**Table 3** The performance of each model on validation set and test set

| Data Set | Evaluation Metrics | CNN Models | | |
|---|---|---|---|---|
| | | ResNet50 | MobileNetV2 | EfficientNetV2B3 |
| Validation | Accuracy | 0.8033 | 0.4133 | 0.7967 |
| | Loss | 0.9451 | 1.9880 | 0.6469 |
| | Precision | 0.8185 | 0.5324 | 0.8474 |
| | Recall | 0.7717 | 0.3950 | 0.8250 |
| | F1-score | 0.7670 | 0.3748 | 0.8231 |
| Test | Accuracy | 0.7354 | 0.3771 | 0.8094 |
| | Loss | 1.1345 | 2.1544 | 0.5829 |
| | Precision | 0.8245 | 0.5359 | 0.8606 |
| | Recall | 0.7675 | 0.4258 | 0.8283 |
| | F1-score | 0.7643 | 0.4122 | 0.8301 |

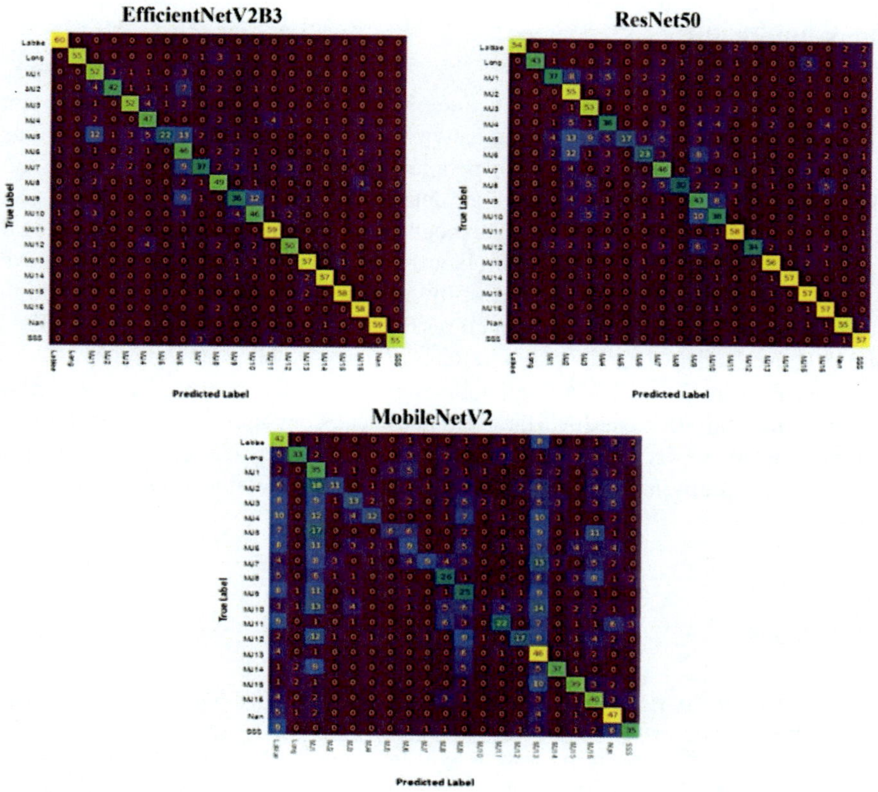


**Fig. 6** Confusion matrices for EfficientNetV2B3, ResNet50 and MobileNetV2 of the test dataset: Abbreviation and name of Pha-tho Teen Jok Fabric Motif are as follows. Lub-lae, Long, MJ1: Chiang saen noi, MJ2: Chiang saen luang, MJ3: Hong pee, MJ4: Hong ploi, MJ5: Khan-sein-sam, MJ6: Khan aew ou, MJ7: Kom roop nok, MJ8: Kom hua mon nai nok non, MJ9: Kud kho bed, MJ10: La khon-klang, MJ11: La khon luang, MJ12: La khon noi, MJ13: Khan sam aew, MJ14: Nak kum, MJ15: Nok kum, MJ16: Nok non kum, Nan, SSS: Si Satchanalai

F1-score and higher test loss, where the model makes the most prediction errors on Pha-tho Teen Jok Mae Jaem fabric motifs. Therefore, it was not a suitable model for classification of Teen Jok Mae Jaem. On the other hand, the results from confusion matrix indicated that MobileNetV2 also not a model to deal with Pha-tho Teen Jok Mae Jaem fabric motif recognition when compared with ResNet50 and EfficientNetV2B3, despite being the fastest to run and with the smallest memory footprint.

## 5 Conclusion

Pha-tho Teen Jok Mae Jaem fabric is one of the commercial products to represent the cultural intellectual wealth and creativity of local Tai people. But with various numbers of motifs, it is difficult for people who are unfamiliar with the Pha-tho Teen Jok fabric motif to recognize their details. However, we showed that we can cast this challenge into a motif image recognition task and used deep convolutional neural network model to tackle the problem. From the results of this study, we have seen that the CNN models can successfully identify Teen Jok Mae Jaem with high accuracy. Specifically, the experimental results suggested that EfficientNetV2B3 is the effective model when compared to ResNet50 and MobileNetV2 models.

Since Pha-tho Teen Jok Mae Jaem fabric motifs are often combined to form a new complex motif, so that one Jok Mae Jaem fabric can contain more than one motif at a time, it would be very interesting to study how to correctly classify those complex motifs concurrently in a similar manner to multi-label classification setting in the future.

## References


1. Chunthone V (2014) The art of contemporary Jok Textiles: the integration between craft and creative design. Veridian E-J 7(4):91–107
2. Burapajatana J (2016) Applying patterns from Identity of Jok woven textiles of Mae Jaem to design of cultural textile products for home decoration. Veridian E-J 9(3):1720–1738
3. Poonsilp K (2018) The development of Thai local woven fabric pattern recognition. Veridian E-J 11(3):2056–2068
4. Girsang ND (2021) Literature study of convolutional neural network algorithm for batik classification. Brilliance Res Artif Intell 1(1):1–7
5. Gultom Y, Arymurthy AM, Masikome RJ (2018) Batik classification using deep convolutional network transfer learning. J Sci Inf 11(2):59–66
6. Puarungroj W, Boonsirisumpun N (2019) Convolutional neural network models for hand-woven fabric motif recognition. In: The 4th international conference on digital arts, media and technology and 2nd ECTI northern section conference on electrical, electronics, computer and telecommunications engineering, pp 300–303
7. Rasyidi MA, Handayani R, Aziz F (2021) Identification of batik making method from images using convolutional neural network with limited amount of data. Bull Electr Eng Inform 10(3):1300–1307
8. Umam C, Krismawan DA, Ali RR (2021) CNN for image identification of hiragana based on pattern recognition. J Appl Intell Syst 6(2):62–71
9. Dwi Putra MT, Kusuma GP (2019) Batik classification using deep learning. Int J Recent Technol Eng (IJRTE) 8(4):11416–11421
10. Rasyidi MA, Bariyah T (2020) Batik pattern recognition using convolutional neural network. Bull Electr Eng Inform (BEEI) 9(4):1430–1437
11. Khodier MM, Ahmed SM, Sayed MS (2022) Complex pattern Jacquard fabrics defect detection using convolutional neural networks and multispectral imaging. IEEE Access 10:10653–10660
12. Negara BS, Satria E, Sanjaya S, Santoso DRD (2021) ResNet-50 for classifying Indonesian batik with data augmentation. In: International congress of advanced technology and engineering (ICOTEN)

13. Yu D, Xu Q, Guo H, Zhao C, Lin Y, Li D (2020) An efficient and lightweight convolutional neural network for remote sensing image scene classification. Sensors 20(7):1–25
14. Tan M, Le QV (2021) EfficientNetV2: smaller models and faster training. In: Proceedings of the 38-th international conference on machine learning, PMLR, vol 139

# Project Management Tools Selection Using BWM TOPSIS

**Piyathep Mahasantipiya and Nuengwong Tuaycharoen**

**Abstract** There are many project management tools on the market. It is arduous to select the best project management tool and meet individual needs. This paper proposes multi-criteria decision-making (MCDM), namely, Best–worst Method (BWM) and Technique for Order Preference by Similarity to Ideal Solution (TOPSIS) to handle the project management tool selection problem and provides criteria for project management tool selection. In our result, the proposed method can assist determine which tool is the best. In addition, we evaluated the method with pairwise matrices. The results revealed that our proposed method is less complicated due to utilizing a smaller number of pairwise comparisons and a smaller number of elements of pairwise comparison matrices, compared with the popular method like the Analytical Hierarchical Process (AHP). And, it is more reliable according to the consistency ratio (CR) being closer to 0 than the AHP in all matrices, i.e., the CR of the pairwise matrix of the criteria with BWM TOPSIS was 0.0157, while AHP was 0.0610.

**Keywords** Project Management Tool · Multi-criteria Decision Making · Best–worst Method · TOPSIS · Analytical Hierarchical Process

## 1 Introduction

Appropriate project management tools and techniques are very significant to project management. They obtain inputs to produce outputs in all project management process groups [1], which can manage the project life cycle, as well as supports Work From Home or Work From Anywhere. Nowadays, there have been many project management tools on the market such as Jira, Monday.com, Asana, Trello, Microsoft (MS) Project, etc. Each project management tool has disparate features.

P. Mahasantipiya (✉) · N. Tuaycharoen
Department of Computer Engineering, Chulalongkorn University, Bangkok, Thailand
e-mail: piyathep.m@student.chula.ac.th

N. Tuaycharoen
e-mail: nuengwong.t@chula.ac.th



P. Meesad et al. (eds.), *Proceedings of the 19th International Conference on Computing and Information Technology (IC2IT 2023)*, Lecture Notes in Networks and Systems 679,
https://doi.org/10.1007/978-3-031-30474-3_14

Most providers have illustrated only the advantages of their tools over others and have defined criteria for project management tool selection to match their tools. These biases have provided only a partial picture for those who intend to utilize the project management tool, not only the project manager but also project team members and stakeholders. Thus, it has been arduous to select the appropriate tool and identify the appropriate criteria for selection.

We studied several studies and found that Norita Ahmad and Phillip A. Laplante [2] presented the software project management tool selection model using AHP, one of the MCDM methods, to handle the software project management tool selection problem. Their model can rank project management tools, leading to the ability to select the appropriate project management tool. Moreover, we found many MCDM methods to handle various decision problems. Therefore, we studied several studies that compared MCDM methods [3–6] to find the most effective and reliable method.

In this paper, we proposed BWM and TOPSIS to handle the project management tools selection problem because BWM is a novel method and requires less comparative effort compared to the most popular methods like AHP [5, 6]. In addition, TOPSIS does not utilize pairwise comparison in the ranking process, leading to less comparative effort. TOPSIS is a simple and intuitive process, and it provides greater agility than any other MCDM method [5]. Both BWM and TOPSIS have been more efficient and reliable than AHP. This work provides appropriate criteria for project management tool selection covering project management knowledge areas from the project management body of knowledge (PMBOK), 6th edition, proposed by the Project Management Institute (PMI). Moreover, we also evaluated our proposed method with computational complexity and consistency to compare the pairwise matrices with AHP to illustrate the efficiency and reliability of BWM TOPSIS as well.

The rest of the paper is structured as follows. Section 2 discusses the related work. Section 3 presents the proposed method. Section 4 discusses our experimental results. Section 5 concludes the paper.

## 2 Related Work

### *2.1 Project Management Knowledge Areas*

Project management knowledge areas are knowledge areas that describe processes, practices, inputs, outputs, tools, and techniques utilized in the project management process proposed by PMI [1]. They are categorized into 10 areas: integration management, scope management, schedule management, cost management, quality management, resource management, communications management, risk management, procurement management, and stakeholder management. The perfect project control and management system should cover project management knowledge areas [7].

## 2.2 Multi-criteria Decision-Making (MCDM)

MCDM is one of the branches of decision-making that has several methods such as AHP, ELimination Et Choix TRaduisant La REalit (ELECTRE), TOPSIS, etc. [8] Each method is very disparate, but usually has the same components such as alternatives, multiple attributes, standard conflicts, unreasonable units, decision weights, and decision matrices. The MCDM has a procedure that involved the numerical analysis of alternatives: determine criteria and alternatives, form numerical measures to the relative importance of the criteria and the impacts of the alternatives for these criteria, and compute the numerical values for alternative ranking. In addition, most MCDMs utilize pairwise comparisons. If there are many pairwise comparisons, it will be more complex and time-consuming [9].

## 2.3 Best-Worst Method (BWM)

BWM is a novel MCDM method developed by Rezaei that weights pairwise comparisons in a way disparate from other methods [10]. Pairwise comparisons of BWM have compared only the best criterion over all other criteria and have compared all other criteria over the worst criterion. Therefore, the number of pairwise comparisons is $2n-3$, and n is the number of criteria. This is a very small number of pairwise comparisons compared to other methods. The steps of BWM are described in Sect. 3.

## 2.4 Technique for Order Preference by Similarity to Ideal Solution (TOPSIS)

TOPSIS is a technique based on the concept that the chosen alternative should have the shortest distance from the ideal solution and the farthest distance from the negative-ideal solution [11]. TOPSIS was developed in 1981 by Yoon and Hwang. It has taken the decision matrix containing alternatives associated with criteria to evaluate ideal solutions and negative-ideal solutions to compute relative closeness, leading to alternative ranking. The steps of TOPSIS are described in Sect. 3.

## 2.5 Analytical Hierarchical Process (AHP)

AHP is the MCDM method developed by Saaty [12]. It has characterized by a hierarchical structure that results in an explicit level of analysis. The AHP hierarchical model consists of 3 levels: Level 1 Goal, Level 2 Criteria and Sub-criteria (if any), and Level 3 Alternatives. Pairwise comparisons of AHP compare all criteria against all criteria. Therefore, the number of pairwise comparisons is n(n−1) ÷ 2, constructing the large comparison effort. The steps of AHP are described in detail in [2, 13].

## 3 Methodology

This paper presents BWM TOPSIS to handle the project management tool selection problem, an overview of all steps is shown in Fig. 1.

### *3.1 Determine Decision Criteria*

We determined the criteria for project management tool selection with the following sub-steps.

1. Criteria survey: We surveyed criteria from various reference sources, which categorized them into 4 websites [14–17] that provide recommendations on project management tools, including the necessary criteria and features, and 4 studies [2, 7, 18, 19] on project management tool selection.
2. Matching with the project management knowledge areas: We matched the criteria and features obtained from our survey with project management knowledge areas from PMBOK, $6^{th}$ edition, to find out which tools or techniques are mentioned or recommended by which areas. This revealed the criteria covering the project management knowledge areas.
3. Expert interview: We interviewed 3 experts who had over 60 days of experience in project management tools for their opinions on the criteria, and we asked in terms of other criteria that should be considered.

Then, we commenced step 1 of BWM, defining criteria as $\{C_1, C_2, \ldots, C_n\}$, n is the number of criteria.

### *3.2 Determine the Preference of the Criteria*

In this step, the pairwise matrix for comparing the best criterion over all other criteria (Best-to-Others) can be constructed as follows.

$$BO = \begin{array}{c} \\ C_B \end{array} \begin{array}{c} C_1 \; C_2 \cdots C_j \cdots C_n \\ \left[ p_{B1} \; p_{B2} \cdots p_{Bj} \cdots p_{Bn} \right] \end{array} \tag{1}$$

The pairwise matrix for comparing all the other criteria over the worst criterion (Others-to-Worst) can be constructed as follows.

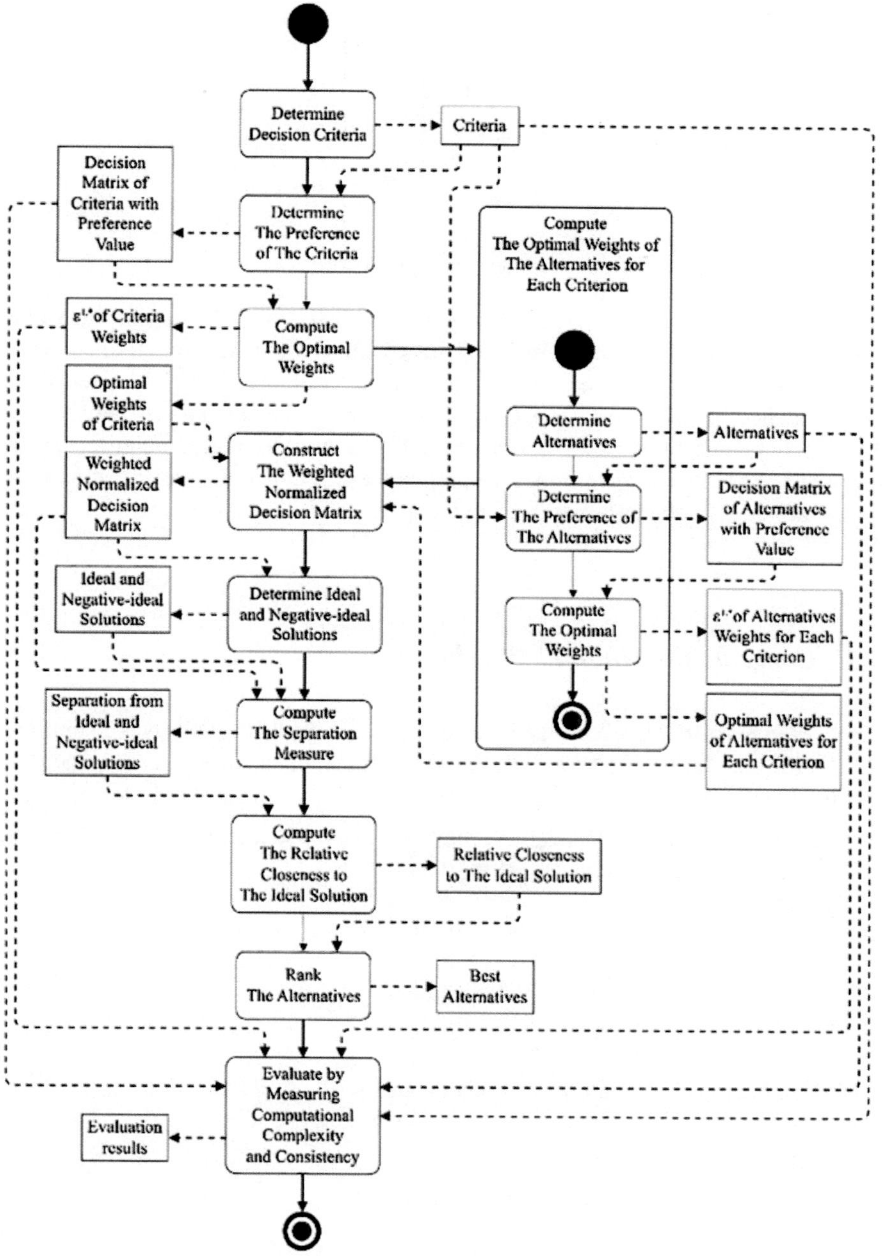


**Fig. 1** Overview of all steps

$$OW = \begin{matrix} \\ C_1 \\ C_2 \\ \vdots \\ C_i \\ \vdots \\ C_n \end{matrix} \begin{matrix} C_W \\ \begin{bmatrix} p_{1W} \\ p_{2W} \\ \vdots \\ p_{iW} \\ \vdots \\ p_{nW} \end{bmatrix} \end{matrix} \quad (2)$$

We performed step 2 of BWM, determining the best criterion $C_B$ and the worst criterion $C_W$ based on decision-maker (DM) opinion. The DM refers to those who intend to utilize the project management tool. In step 3 of BWM, the DM determines the preferences for Best-to-Others comparison $(p_{Bj})$ utilizing a number between 1–9 where 1 is most desirable or significant, and 9 is least desirable or significant. In step 4 of BWM, the DM determines the preferences for Others-to-Worst comparison $(p_{jW})$ utilizing a number between 1–9 where 1 is least desirable or significant, and 9 is most desirable or significant.

### 3.3 Compute the Optimal Weights

This is step 5 of BWM. We utilized the equation of Linear BWM [20]. The optimal weight for comparing the best criterion over all the other criteria is the one that satisfies the condition $\frac{w_B}{w_j} = p_{Bj}$ and for comparing all the other criteria over the worst criterion is the one that satisfies the condition $\frac{w_j}{w_W} = p_{jW}$ for all $j$. The conditions can be formulated as follows.

$$\begin{aligned} &min\ \ max_j\{|w_B - p_{Bj}w_j|, |w_j - p_{jW}w_W|\} \\ &\text{s.t.}\ \sum\nolimits_j w_j = 1 \\ &w_j \geq 0, \forall j \end{aligned} \quad (3)$$

(3) can be transferred into the following equation.

$$\begin{aligned} &min\ \varepsilon^L \\ &|w_B - p_{Bj}w_j| \leq \varepsilon^L, \forall j \\ &\text{s.t.}\ |w_j - p_{jW}w_W| \leq \varepsilon^L, \forall j \\ &\sum\nolimits_j w_j = 1 \\ &w_j \geq 0, \forall j \end{aligned} \quad (4)$$

Solving (4) yields the optimal weight $(w_1^*, w_2^*, \ldots, w_n^*)$ and $\varepsilon^{L*}$.

### 3.4 Compute the Optimal Weights of the Alternatives for Each Criterion

We determined the alternatives (hereinafter referred to as project management tools) the DM intended to rank $\{A_1, A_2, \ldots, A_m\}$, m is the number of project management tools, and we determined preferences of project management tools over each criterion based on the DM's opinion, then we computed the weights like in the previous step, but performed it with the project management tools instead of the criteria. This is still the step of BWM. Therefore, the optimal weights of the alternatives for each criterion are as follows.

$$D = \begin{array}{c} \\ A_1 \\ A_2 \\ \vdots \\ A_i \\ \vdots \\ A_m \end{array} \begin{array}{c} \begin{array}{cccccc} C_1 & C_2 & \cdots & C_j & \cdots & C_n \end{array} \\ \begin{bmatrix} t_{11} & t_{12} & \cdots & t_{1j} & \cdots & t_{1n} \\ t_{21} & t_{22} & \cdots & t_{2j} & \cdots & t_{2n} \\ \vdots & \vdots & & \vdots & & \vdots \\ t_{i1} & t_{i2} & \cdots & t_{ij} & \cdots & t_{in} \\ \vdots & \vdots & & \vdots & & \vdots \\ t_{m1} & t_{m2} & \cdots & t_{mj} & \cdots & t_{mn} \end{bmatrix} \end{array} \tag{5}$$

### 3.5 Construct the Weighted Normalized Decision Matrix

From this step onwards, we commenced step 1 of TOPSIS, normalizing each optimal weight of the project management tools for each criterion $t_{ij}$ as follows.

$$z_{ij} = \frac{t_{ij}}{\sqrt{\sum_{i=1}^{n} t_{ij}^2}} \tag{6}$$

Then we performed step 2 of TOPSIS, multiplying each normalized decision matrix a$z_{ij}$ with each criteria weight $w_j$ to construct the weighted normalized decision matrix $v_{ij}$ as follows.

$$v_{ij} = w_j z_{ij} \tag{7}$$

### 3.6 Determine Ideal and Negative-Ideal Solutions

Step 3 of TOPSIS is determining the ideal solution d negative-ideal solution utilizing the min and max values that can be found from the weighted normalized decision matrix $v_{ij}$. as follows.

$$
\begin{aligned}
a_j^+ &= max\{v_{1j}, v_{2j}, \ldots, v_{ij}, \ldots, v_{mj}\} \\
a_j^- &= min\{v_{1j}, v_{2j}, \ldots, v_{ij}, \ldots, v_{mj}\} \\
A^+ &= \begin{bmatrix} a_1^+ & a_2^+ & \cdots & a_j^+ & \cdots & a_n^+ \end{bmatrix} \\
A^- &= \begin{bmatrix} a_1^- & a_2^- & \cdots & a_j^- & \cdots & a_n^- \end{bmatrix}
\end{aligned}
\tag{8}
$$

### 3.7 Compute the Separation Measure

In Step 4 of TOPSIS, the separation from the ideal solution $S_i^+$ and the separation from the negative-ideal solution $S_i^-$ can be computed utilizing the weighted normalized decision matrix $v_{ij}$, ideal solution $a_j^+$, and negative-ideal solution $a_j^-$ as follows.

$$
\begin{aligned}
S_i^+ &= \sqrt{\sum\nolimits_{j=1}^{n} \left(v_{ij} a_j^+\right)^2} \\
S_i^- &= \sqrt{\sum\nolimits_{j=1}^{n} \left(v_{ij} - a_j^-\right)^2}
\end{aligned}
\tag{9}
$$

### 3.8 Compute the Relative Closeness to the Ideal Solution

In Step 5 of TOPSIS, the relative closeness of each alternative $r_i$ can be computed utilizing the separation from the ideal solution $S_i^+$ and the separation from the negative-ideal solution $S_i^-$ as follows.

$$
r_i = \frac{S_i^-}{S_i^+ + S_i^-} \tag{10}
$$

### 3.9 Rank the Alternatives

This is step 6 and the last step of TOPSIS. The project management tools can be ranked in descending order of the relative closeness $r_i$. The highest relative closeness indicates the best project management tool.

### 3.10 Evaluate by Measuring Computational Complexity and Consistency

We measured computational complexity to recognize efficiency and measured consistency to recognize reliability. The measurement process is as follows.

- Computational Complexity: Given $n$ is the number of criteria, and $m$ is the number of project management tools. In BWM TOPSIS, the number of elements in the pairwise comparison matrices is $2n + n(2m)$, and the number of pairwise comparisons is $2n-3 + n(2m-3)$. In AHP, the number of elements in the pairwise comparison matrices is $1(n \times n) + n(m \times m)$, and the number of pairwise comparisons is $\frac{n(n-1)+n \times m(m-1)}{2}$.
- Consistency: In this section, we computed the consistency ratio ($CR$). In BWM, Rezaei presented a computational method utilizing $\varepsilon^{L*}$ derived from (4) and $max_\varepsilon$ or the consistency index ($CI$) varies with $p_{BW}$ as shown in Table 1, then compute $CR$ as follows.

$$CR = \frac{\varepsilon^{L*}}{CI} \tag{11}$$

In AHP, $CR$ can be computed utilizing $\lambda_{max}$ that can be computed from the pairwise matrix of AHP and the number of criteria $n$ (or the number of project management tools $m$) to compute $CI$ as follows.

$$CI = \frac{\lambda_{max} - n}{n - 1} \tag{12}$$

Then computed the random consistency index ($RI$) varies with $n$ (or $m$) as shown in Table 2 as follows.

$$CR = \frac{CI}{RI} \tag{13}$$

**Table 1** Consistency Index of BWM [10]

| $p_{BW}$ | 1 | 2 | 3 | 4 | 5 | 6 | 7 | 8 | 9 |
|---|---|---|---|---|---|---|---|---|---|
| $CI$ ($max_\varepsilon$) | 0.00 | 0.44 | 1.00 | 1.63 | 2.30 | 3.00 | 3.73 | 4.47 | 5.23 |

**Table 2** Random consistency Index [12]

| $n$ | 3 | 4 | 5 | 6 | 7 | 8 | 9 | 10 | 11 | 12 |
|---|---|---|---|---|---|---|---|---|---|---|
| $RI$ | 0.52 | 0.89 | 1.12 | 1.26 | 1.36 | 1.41 | 1.46 | 1.49 | 1.52 | 1.54 |

# 4 Experimental Result

From the proposed method, we have obtained 4 main experimental results.

## 4.1 Criteria for Project Management Tool Selection

The criteria obtained from our survey and matched to project management knowledge areas are shown in Table 3, and the results of interviews with 3 experts are shown in Table 4. In addition, the experts recommended criteria that should be considered: integration, price, and ease of use. Therefore, there are 11 criteria for project management tool selection as shown in Table 5. Based on the DM opinion, task management $C_3$ was selected as the best criterion $C_B$, and team collaboration $C_4$ was selected as the worst criterion $C_W$. Table 6 shows the preference for $C_B$-to-$C_j$ based on DM opinion. Table 7 shows the preference for $C_j$-to-$C_W$ based on DM opinion. The optimal weights of criteria are computed utilizing (3) and (4), resulting in criteria weights $w_j$ as shown in Table 8, and $\varepsilon^{L*}$ of the pairwise matrix of criteria equals 0.0822.

**Table 3** Criteria and PMBOK 6th's Project Management Knowledge Areas Matching

| Project management knowledge areas | Criteria | | | | | | | |
|---|---|---|---|---|---|---|---|---|
| | Planning and Scheduling | Workload and Resource Management | Task Management | Team Collaboration | Dashboard | Security | Document Management | Mobile Version |
| Integration Management | Gantt Chart (or Bar Chart or WBS), and Project Calendar | | | Share Centralized Data | Dashboard | Access Control | Document Control | |
| Scope Management | Gantt Chart (or Bar Chart or WBS), and Project Requirements | | | Feedback | | | Document Control | |
| Schedule Management | Gantt Chart (or Bar Chart or WBS), Roadmap, and Project Calendar | | Kanban Board | Share Centralized Data | | | Document Control | |
| Cost Management | Gantt Chart (or Bar Chart or WBS) | Cost Tracking | | | | | Document Control | |
| Quality Management | Project Requirements | | | | | | Document Control | |

(continued)

**Table 3** (continued)

| Project management knowledge areas | Criteria | | | | | | | |
|---|---|---|---|---|---|---|---|---|
| | Planning and Scheduling | Workload and Resource Management | Task Management | Team Collaboration | Dashboard | Security | Document Management | Mobile Version |
| Resource Management | Gantt Chart (or Bar Chart or WBS) | | | Share Centralized Data, Feedback, and Email | | | Document Control | |
| Communications Management | | | | Share Centralized Data, Feedback, Email, and Messaging (or Chat) | Dashboard | Access Control | Document Control | |
| Risk Management | Gantt Chart (or Bar Chart or WBS), and Project Requirements | Cost Tracking | | | | | Document Control | |
| Procurement Management | Project Requirements | | | | | | Document Control | |
| Stakeholder Management | Project Requirements | | | Share Centralized Data, and Feedback | | Access Control | Document Control | |

**Table 4** Expert interview results

| Criteria | Expert 1 | Expert 2 | Expert 3 |
|---|---|---|---|
| Planning and Scheduling | Agree | Agree | Agree |
| Workload and Resource Management | Agree | Agree | Agree |
| Task Management | Agree | Agree | Agree |
| Team Collaboration | Disagree | Agree | Agree |
| Dashboard | Agree | Agree | Agree |
| Security | Disagree | Agree | Agree |
| Document Management | Agree | Agree | Agree |
| Mobile Version | Disagree | Agree | Agree |

**Table 5** Criteria for project management tool selection

| Symbol | Criterion | Description |
|---|---|---|
| $C_1$ | Planning and Scheduling | This criterion is utilized to evaluate project management tools for goal setting, planning, defining and managing requirements, project progress tracking, task schedule management, and managing people responsible for each task. The relevant features include a Gantt chart (or Bar Chart, or WBS), Roadmap, Project Calendar, and Project Requirement |
| $C_2$ | Workload and Resource Management | This criterion is utilized to evaluate project management tools for managing, evaluating, and tracking project resources, costs, unnecessary expenses, team members, and workloads. The relevant features include Cost Tracking, Workload Management, and Resource Management |
| $C_3$ | Task Management | This criterion is utilized to evaluate project management tools for managing, prioritizing, assigning, resolving, and tracking tasks, including task automatic notification. The relevant features include Kanban Board |
| $C_4$ | Team Collaboration | This criterion is utilized to evaluate project management tools for team collaboration by considering communication, comments, recommendations, conversation tracking, sharing information, documents, contract, and communications that can substitute email or email integration. The relevant features include Messaging (or Chat), Document Sharing, Commenting, and Email Integration |
| $C_5$ | Dashboard | This criterion is utilized to evaluate project management tools for displaying data and relevant indicators. The dashboard should be able to display information in an easily interpretable format, easily share project details with teams or customers, and be customizable according to each stakeholder's needs |
| $C_6$ | Security | This criterion is utilized to evaluate project management tools for ensuring the security of the system is still reliable and free from intrusion. The relevant features include Access Control, Authentication Method, Security Certificate, and Encryption |
| $C_7$ | Document Management | This criterion is utilized to evaluate project management tools for facilitating document management, version control, change tracking, document history viewing, and automatic file backup. The relevant features include Document Control and Document Repository |
| $C_8$ | Mobile Version | This criterion is utilized to evaluate project management tools for the ability to manage projects anytime, anywhere through the mobile application |

(continued)

**Table 5** (continued)

| Symbol | Criterion | Description |
|---|---|---|
| $C_9$ | Integration | This criterion is utilized to evaluate project management tools for integration with other applications, which makes project management tools more capable and able to take advantage of the integrated applications |
| $C_{10}$ | Price | This criterion is utilized to evaluate project management tools for the price. Price should be computed for the same currency, the same number of users, and the same period. For example, $10/user/month |
| $C_{11}$ | Ease of Use | This criterion is utilized to evaluate project management tools for ease of use. Several tools offer free trials. Users can make an effort to utilize the various functions preceding making a purchase |

**Table 6** Preference of best-to-others of criteria

| $BO$ | $C_1$ | $C_2$ | $C_3$ | $C_4$ | $C_5$ | $C_6$ | $C_7$ | $C_8$ | $C_9$ | $C_{10}$ | $C_{11}$ |
|---|---|---|---|---|---|---|---|---|---|---|---|
| $C_3$ | 2 | 3 | 1 | 9 | 4 | 7 | 6 | 8 | 8 | 5 | 8 |

**Table 7** Preference of others-to-worst of criteria

| $OW$ | $C_4$ |
|---|---|
| $C_1$ | 8 |
| $C_2$ | 8 |
| $C_3$ | 9 |
| $C_4$ | 1 |
| $C_5$ | 8 |
| $C_6$ | 5 |
| $C_7$ | 6 |
| $C_8$ | 3 |
| $C_9$ | 4 |
| $C_{10}$ | 7 |
| $C_{11}$ | 2 |

**Table 8** Criteria weights

| Criterion | Weight | Criterion | Weight | Criterion | Weight | Criterion | Weight |
|---|---|---|---|---|---|---|---|
| $C_1$ | 0.1787 | $C_4$ | 0.0214 | $C_7$ | 0.0596 | $C_{10}$ | 0.0715 |
| $C_2$ | 0.1191 | $C_5$ | 0.0894 | $C_8$ | 0.0447 | $C_{11}$ | 0.0447 |
| $C_3$ | 0.2752 | $C_6$ | 0.0511 | $C_9$ | 0.0447 | | |

## 4.2 The Best Project Management Tool

To evaluate the project management tools, the DM determined the tools, including Jira $A_1$, Trello $A_2$, and MS Project $A_3$, for example. Based on the DM opinion, Table 9 shows the preference for $A_B$-to-$A_j$ for each criterion $t_{Bj}$, for example, column 1 represents the preference for $A_B$-to-$A_j$ for $C_1$. Similarly, Table 10 shows the preference for $A_j$-to-$A_W$ for each criterion $t_{iW}$, for example, column 1 represents the preference for $A_j$-to-$A_W$ for $C_1$. The optimal weights of the project management tools for each criterion $t_{ij}$ were computed utilizing (3) and (4) as shown in Table 11 and $\varepsilon^{L*}$ of the pairwise matrix of the project management tools for each criterion $A_j$ - to - $C_j$ as shown in Table 16. Afterward, the weighted normalized decision matrix $v_{ij}$ was constructed utilizing (6) and (7) as shown in Table 12. The ideal solution $a_j^+$ and negative-ideal solution $a_j^-$ were determined utilizing (8) as shown in Table 13. The separation from the ideal solution $S_i^+$ and the separation from the negative-ideal solution $S_i^-$ was computed utilizing (9), and the relative closeness of each project management tool $r_i$ was computed utilizing (10) as shown in Table 14. Moreover, Table 14 represents the ranked project management tools utilizing the relative closeness, resulting in rank 1 or the best project management tool is Jira $A_1$, because it has the highest relative closeness.

**Table 9** Preference of best-to-others of the project management tools for each criterion

| *BO* | $C_1$ | $C_2$ | $C_3$ | $C_4$ | $C_5$ | $C_6$ | $C_7$ | $C_8$ | $C_9$ | $C_{10}$ | $C_{11}$ |
|---|---|---|---|---|---|---|---|---|---|---|---|
| $A_1$ | 1 | 1 | 1 | 3 | 1 | 2 | 1 | 2 | 1 | 5 | 5 |
| $A_2$ | 3 | 5 | 5 | 1 | 5 | 1 | 2 | 1 | 2 | 1 | 1 |
| $A_3$ | 9 | 9 | 9 | 9 | 9 | 9 | 9 | 9 | 9 | 9 | 9 |

**Table 10** Preference of others-to-worst of the project management tools for each criterion

| *OW* | $C_1$ | $C_2$ | $C_3$ | $C_4$ | $C_5$ | $C_6$ | $C_7$ | $C_8$ | $C_9$ | $C_{10}$ | $C_{11}$ |
|---|---|---|---|---|---|---|---|---|---|---|---|
| $A_1$ | 9 | 9 | 9 | 7 | 9 | 8 | 9 | 8 | 9 | 5 | 5 |
| $A_2$ | 5 | 5 | 5 | 9 | 5 | 9 | 8 | 9 | 8 | 9 | 9 |
| $A_3$ | 1 | 1 | 1 | 1 | 1 | 1 | 1 | 1 | 1 | 1 | 1 |

**Table 11** The weights of the project management tools for each criterion

| | $C_1$ | $C_2$ | $C_3$ | $C_4$ | $C_5$ | $C_6$ | $C_7$ | $C_8$ | $C_9$ | $C_{10}$ | $C_{11}$ |
|---|---|---|---|---|---|---|---|---|---|---|---|
| $A_1$ | 0.6800 | 0.7524 | 0.7524 | 0.2706 | 0.7524 | 0.3472 | 0.5972 | 0.3472 | 0.5972 | 0.1810 | 0.1810 |
| $A_2$ | 0.2533 | 0.1810 | 0.1810 | 0.6706 | 0.1810 | 0.5972 | 0.3472 | 0.5972 | 0.3472 | 0.7524 | 0.7524 |
| $A_3$ | 0.0667 | 0.0667 | 0.0667 | 0.0588 | 0.0667 | 0.0556 | 0.0556 | 0.0556 | 0.0556 | 0.0667 | 0.0667 |

**Table 12** The weighted normalized decision matrix

| | $C_1$ | $C_2$ | $C_3$ | $C_4$ | $C_5$ | $C_6$ | $C_7$ | $C_8$ | $C_9$ | $C_{10}$ | $C_{11}$ |
|---|---|---|---|---|---|---|---|---|---|---|---|
| $A_1$ | 0.1215 | 0.0896 | 0.2071 | 0.0058 | 0.0672 | 0.0177 | 0.0356 | 0.0155 | 0.0267 | 0.0129 | 0.0081 |
| $A_2$ | 0.0453 | 0.0216 | 0.0498 | 0.0144 | 0.0162 | 0.0305 | 0.0207 | 0.0267 | 0.0155 | 0.0538 | 0.0336 |
| $A_3$ | 0.0119 | 0.0079 | 0.0183 | 0.0013 | 0.0060 | 0.0028 | 0.0033 | 0.0025 | 0.0025 | 0.0048 | 0.0030 |

**Table 13** Ideal and negative-ideal solutions

| | $C_1$ | $C_2$ | $C_3$ | $C_4$ | $C_5$ | $C_6$ | $C_7$ | $C_8$ | $C_9$ | $C_{10}$ | $C_{11}$ |
|---|---|---|---|---|---|---|---|---|---|---|---|
| $A^+$ | 0.1215 | 0.0896 | 0.2071 | 0.0144 | 0.0672 | 0.0305 | 0.0356 | 0.0267 | 0.0267 | 0.0538 | 0.0336 |
| $A^-$ | 0.0119 | 0.0079 | 0.0183 | 0.0013 | 0.0060 | 0.0028 | 0.0033 | 0.0025 | 0.0025 | 0.0048 | 0.0030 |

**Table 14** Separation from the ideal and negative-ideal solution, relative closeness, and ranking of the project management tools

| | $S_i^+$ | $S_i^-$ | $r_i$ | Rank |
|---|---|---|---|---|
| $A_1$ | 0.0518 | 0.2453 | 0.8257 | 1 |
| $A_2$ | 0.1953 | 0.0879 | 0.3104 | 2 |
| $A_3$ | 0.2541 | 0.0000 | 0.0000 | 3 |

## *4.3 Computational Complexity*

The computational complexity we evaluated is shown in Table 15. This reveals that BWM TOPSIS is more efficient than AHP because BWM TOPSIS requires $2 \times 11 + 11 \times (2 \times 3) = 88$ elements in pairwise comparison matrices while AHP requires $1 \times (11 \times 11) + 11 \times (3 \times 3) = 220$ elements in pairwise comparison matrices and BWM TOPSIS requires $2 \times 11 - 3 + 11 \times (2 \times 3 - 3) = 52$ pairwise comparisons while AHP requires $\frac{11 \times (11 - 1) + 11 \times 3 \times (3 - 1)}{2} = 88$ pairwise comparisons.

**Table 15** Comparison of the computational complexity of BWM TOPSIS vs. AHP

| | | |
|---|---|---|
| Number of criteria | 11 | |
| Number of the project management tools | 3 | |
| Method | BWM TOPSIS | AHP |
| Number of elements in pairwise comparison matrices | 88 | 220 |
| Number of pairwise comparisons | 52 | 88 |

**Table 16** Comparison of the consistency of BWM TOPSIS vs. AHP

| Pairwise Matrix | Method | | | | | | |
|---|---|---|---|---|---|---|---|
| | BWM TOPSIS | | | AHP | | | |
| | $\varepsilon^{L*}$ | $CI$ | $CR$ | $\lambda_{max}$ | $CI$ | $RI$ | $CR$ |
| $C_j$ | 0.0822 | 5.23 | 0.0157 | 11.9264 | 0.0926 | 1.52 | 0.0610 |
| $A_j$-to-$C_1$ | 0.0800 | 5.23 | 0.0153 | 3.0446 | 0.0223 | 0.52 | 0.0429 |
| $A_j$-to-$C_2$ | 0.1524 | 5.23 | 0.0291 | 3.2034 | 0.1017 | 0.52 | 0.1956 |
| $A_j$-to-$C_3$ | 0.1524 | 5.23 | 0.0291 | 3.2034 | 0.1017 | 0.52 | 0.1956 |
| $A_j$-to-$C_4$ | 0.1412 | 5.23 | 0.0270 | 3.1217 | 0.0608 | 0.52 | 0.1170 |
| $A_j$-to-$C_5$ | 0.1524 | 5.23 | 0.0291 | 3.2034 | 0.1017 | 0.52 | 0.1956 |
| $A_j$-to-$C_6$ | 0.0972 | 5.23 | 0.0186 | 3.0518 | 0.0259 | 0.52 | 0.0499 |
| $A_j$-to-$C_7$ | 0.0972 | 5.23 | 0.0186 | 3.0518 | 0.0259 | 0.52 | 0.0499 |
| $A_j$-to-$C_8$ | 0.0972 | 5.23 | 0.0186 | 3.0518 | 0.0259 | 0.52 | 0.0499 |
| $A_j$-to-$C_9$ | 0.0972 | 5.23 | 0.0186 | 3.0518 | 0.0259 | 0.52 | 0.0499 |
| $A_j$-to-$C_{10}$ | 0.1524 | 5.23 | 0.0291 | 3.2034 | 0.1017 | 0.52 | 0.1956 |
| $A_j$-to-$C_{11}$ | 0.1524 | 5.23 | 0.0291 | 3.2034 | 0.1017 | 0.52 | 0.1956 |

### 4.4 Consistency

CR ranges between 0 and 1. CR close to 0 indicates more consistency, while CR close to 1 indicates less consistency. The consistency we evaluated is shown in Table 16. This revealed that BWM TOPSIS is more reliable than AHP because the CR of BWM is closer to 0 than AHP in both criteria $C_j$ comparison and all project management tools for each criterion $A_j$ - to - $C_j$ comparison.

## 5 Conclusion

In this paper, we proposed 11 criteria for project management tool selection that cover the project management knowledge areas from PMBOK, 6th edition, proposed by PMI, and our criteria are also accepted by experts who had over 60 days of experience in project management tools. We propose BWM TOPSIS to handle the project management tool selection problem. The criteria weights and decision matrix were obtained by performing BWM. Project management tools were ranked by performing TOPSIS, which reveals the best project management tool. Our evaluation results obtained by computing the computational complexity and consistency revealed that BWM TOPSIS is more efficient and reliable than popular methods like AHP. However, we hope our work will greatly benefit those who show enthusiasm regarding BWM TOPSIS or an appropriate method for project management tool selection.

## References


1. A guide to the project management body of knowledge (PMBOK Guide) 6th edn. Project Management Institute (2017)
2. Ahmad N, Laplante PA (2008) software project management tools: making a practical decision Using AHP, In: 30th Annual IEEE/nasa software engineering workshop, pp 76–84, Columbia, MD, USA
3. Sabilla Ajrina A, Sarno R, Hari Ginardi RV (2018) Comparison of AHP and BWM methods based on geographic information system for determining potential zone of Pasir Batu mining, In: International seminar on application for technology of information and communication, pp 453–457, Semarang, Indonesia
4. Alrababah SAA, Gan KH, Tan TP (2017) Comparative analysis of MCDM methods for product aspect ranking: TOPSIS and VIKOR, In: 8th international conference on information and communication systems (ICICS), pp 76–81, Irbid, Jordan
5. Youssef AE (2020) An integrated MCDM approach for cloud service selection based on TOPSIS and BWM. IEEE Access 8:71851–71865
6. Fernandes JM, Prozil Rodrigues S, Costa LA (2015) Comparing AHP and ELECTRE I for prioritizing software requirements, In: IEEE/ACIS 16th international conference on software engineering, artificial intelligence, networking and parallel/distributed computing (SNPD), Takamatsu, Japan
7. Lei F, Li S, Ying Y, Bengong Y.: The selection of project management software by FAHP and FMCDM in automobile R&D process, In: International conference on networking and digital society, pp 66–69
8. Triantaphyllou E (2000) Multi-criteria decision making methods a comparative study: Springer, New York
9. Farshidi S, Jansen S, De Jong R, Brinkkemper S (2011) A Decision support system for cloud service provider selection problem in software producing organizations, In: IEEE 20th conference on business informatics (CBI), pp.139–148, Vienna, Austria
10. Rezaei J (2015) Best-worst multi-criteria decision-making method, Omega 58:49–57
11. Hwang CL, Yoon K (1981) Methods for multiple attribute decision making. In: multiple attribute decision making. LNCMS vol 186. Springer, Berlin. https://doi.org/10.1007/978-3-642-48318-9_3
12. Saaty TL (1980) The analytic hierarchy process (1980)
13. Teknomo K (2020) Analytic hierarchy process (AHP) Tutorial
14. Selecting the best project management software https://www.planview.com/resources/guide/what-is-project-management/selecting-best-project-management-software/. Accessed 22 Dec 2022
15. 10 Criteria for selecting a project management tool, https://www.zentao.pm/blog/criteria-selecting-project-management-tool-1048.html. Accessed 22 Dec 2022
16. Easy Projects and Birdview blog. https://explore.easyprojects.net/blog/project-management-software-requirements. Accessed 22 Dec 2022
17. Considerations for Choosing Project Management Tools. https://project-management.com/10-considerations-for-choosing-a-project-management-tool/. Accessed 22 Dec 2022
18. Muszyńska K, Swacha J (2017) Project management and communication software selection using the weighted regularized HASSE method, In: Proceedings of the 2017 federated conference on computer science and information systems, pp 919–923
19. Shirokova SV, Bolsunovskaya MV, Gintciak AM, Zharova MS (2018) The substantiation of the project management system for company by means of the functional completeness criteria in innovative company, In: Third international conference on human factors in complex technical systems and environments (ERGO)s and Environments (ERGO), St. Petersburg, Russia
20. Rezaei J (2016) Best-worst multi-criteria decision-making method: some properties and a linear model. Omega 64:126–130

# Holistic Evaluation Framework for VR Industrial Training

**Nattamon Srithammee and Prajaks Jitngernmadan**

**Abstract** Effectiveness of Virtual Reality (VR) training in the industrial training is still a controversial issue among industrialists and experts, despite the fact that VR is capable and powerful. To emphasize the ability of industrial VR training, we propose a holistic evaluation framework that proofs effectiveness of using VR technology in industrial training. The framework includes the process of VR content preparation, data collection methods, and data evaluation methods. We collected the trainees' activity data within the VR system and analyzed it with ANCOVA and Cosine Similarity methods. The results showed that the use of VR technology in industrial training could improve the trainees' skills significantly.

**Keywords** Virtual Reality · Industrial Training · Effectiveness · Evaluation · Framework

## 1 Introduction

The concept of Industrial 4.0 is supported by the usage of information and communication technologies from things, including the internet of things, augmented and virtual reality, and machine-to-machine communication, with a focus on industrial automation and machine-to-machine communication. Training using Virtual Reality (VR) technology can also be used to prepare and communicate with the old and same workforce that will work in the new automation industry. One advantage of VR-based training is that it offers a realistic and immersive learning environment that otherwise is difficult to replicate in the real world. This is particularly useful for training in high-risk or dangerous situations, where real-world training can be impractical or dangerous. In industry 4.0, the VR simulates the real feel for the trainees as if they

N. Srithammee (✉) · P. Jitngernmadan
Faculty of Informatics, Burapha University, Chon Buri, Thailand
e-mail: 63910160@go.buu.ac.th

P. Jitngernmadan
e-mail: prajaks@buu.ac.th

 
P. Meesad et al. (eds.), *Proceedings of the 19th International Conference on Computing and Information Technology (IC2IT 2023)*, Lecture Notes in Networks and Systems 679,
https://doi.org/10.1007/978-3-031-30474-3_15

were on the production line in the industry. The benefits include the training scalability, versatile training scenarios, cost efficiency, and data capturing and collecting for later analysis. The VR training will play an increasingly important role in product development by providing unprecedented opportunities for integrating and organizing data, as well as new spaces for knowledge sharing and training. The modern Industry 4.0 paradigm's technological economic expansion has continued to improve the degree of sophistication in terms of product design, innovation, implementation, performance evaluation, and so on. As a result, educational institutions must pay more attention to the development and implementation of educational programs and curricula that use modern teaching methods. Although VR has been mentioned in up-to-date practical training environments at leading universities or training centers, they are not yet implemented properly. One important question is how to ensure that the training with VR technology is as good as the training with traditional physical equipment.

The Virtual Reality (VR) technology is a computer-generated simulation of a 3D environment that can be interacted with in a realistic or physical way by a person using VR equipment, such as a headset or gloves with sensors. VR is often used for entertainment, such as video games, but it also has a wide range of other applications, including education, training, and therapy [1]. In education and training, VR can provide immersive and interactive learning experiences that allow people to learn new skills or concepts through hands-on experiences. For example, VR can be used to simulate real-world scenarios in fields such as Mining, Healthcare, Sports Science, or Industry 4.0 [2–4], allowing people to practice and develop their skills in a safe and controlled environment.

In this research, the effectiveness of VR-based training in the context of industrial automation is evaluated holistically. The conceptual approach for the evaluation possibility is described and the results of the assessment are presented. A framework for evaluating industrial VR-based training is also developed. The training focuses on the Electrical Vehicle Assembly task, which is a part of industrial automation training. This work is organized into five sections, including 1) Introduction, 2) Theories and Related Works, 3) Conceptual Approach, 4) Results and Discussion, and 5) Conclusion and Outlook.

## 2 Theories and Related Works

A literature review was conducted to examine the evaluation methods of using Virtual Reality in the context of Industry 4.0. The review included 60 papers that were published between 2017 and 2022. From the analysis, they are categorized into 3 groups according to the used evaluation methods and concepts as follows:

## *2.1 Pedagogical Approach*

A pedagogical approach refers to the principles and methods used by instructors to facilitate didactical learning. It involves making decisions about the content being taught, the methods of instruction, and the ways in which students are assessed. Some common pedagogical approaches include Constructivism, Project-based learning, Inquiry-based learning, and Problem-based learning. Constructivism emphasizes the active role of the learner in constructing their own understanding of new concepts [5]. Project-based learning involves students working on real-world projects that require the application of knowledge and skills. Inquiry-based learning involves students asking questions and conducting research to explore and understand new concepts. In addition, Problem-based learning involves students working on real or simulated problems that require the application of knowledge and skills to find solutions. These traditional learning concepts can be used to evaluate some specific parts of VR training.

## *2.2 Applied Training Model Approach*

The applied training model approach is a teaching method that focuses on practical, real-world application of knowledge and skills. In the context of Virtual Reality, the applied training model approach might involve the use of VR experiences to teach practical skills or to allow students to practice and apply what they have learned in a simulated environment. The ADDIE model [6] and Kirkpatrick model [7] are used in the virtual reality training program for healthcare professionals is an example of the applied training model approach in VR. This program used VR simulations to teach practical skills, such as wound care and medication administration, allowing healthcare students to practice and apply their knowledge in a simulated clinical environment [8]. Quality of experience (QoE) metrics can also be used to evaluate the effectiveness of VR applications [9]. QoE refers to the overall satisfaction and enjoyment of a user when interacting with a system or service and can be used to assess the usefulness and effectiveness of VR experiences.

## *2.3 Biometric Measurement Approach*

Biometric measurement refers to the use of physiological or behavioral characteristics to identify and authenticate individuals. In the context of Virtual Reality, biometric measurement can be used to assess and measure the impact of virtual experiences on individuals [10]. This can be done using a variety of biometric data, such as brain waves, heart rate, and eye movements. These can be measured using specialized technology such as electroencephalography (EEG) machines [11] and eye-tracking devices.

The mentioned approaches above only tackle some parts of the evaluation problem in VR. We aimed to design and develop a holistic evaluation framework that can construct and evaluation a hands-on industrial training using VR technology.

# 3 Conceptual Approach

Virtual Reality technology has the potential to enhance learning and increase student engagement by providing immersive, interactive experiences that capture students' attention and make learning more enjoyable. One of the most important questions is how to measure the effectiveness of the VR technology for hands-on industrial trainings. With this question, we constructed the approach for a holistic evaluation of VR.

## 3.1 Approach Overview

The holistic evaluation framework must see the industrial training using VR technology as a whole. This means that the process has to be monitored at the beginning, namely the content development. We construct our approach in 3 phases, including Phase I Development, Phase II Evaluation, and Phase II Framework Development, whereas Phase I dedicates to content development, Phase II dedicates to analysis and evaluation according to statistical approaches, and Phase III dedicates to the Framework development. Figure 1 depicts the overview of our approach.

The Fig. 1 shows the overview of our approach. In the Phase I, the necessary contents were created under the ADDIE approach, which is a well-known pedagogic approach. The industrial VR training system consists of 3 main modules that are Learning module, Training module, and Testing module. This training environment was created and simulated using 3D virtual objects, and therefore they can create

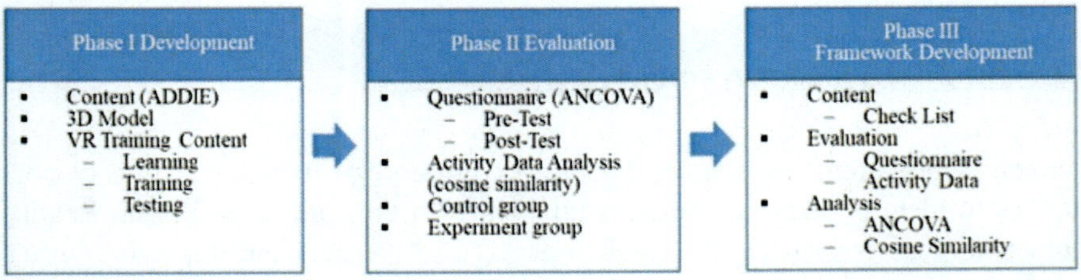


**Fig. 1** Approach overview of the holistic evaluation framework

the realistic training experiences virtually. In our experiment, we recruited 10 engineering students from a Mechatronic department, where the students are familiar with industrial automation and related. These students were then separated into a treatment group and a control group, 5 students per group. The control group watched the training videos, where the students of the treatment group used the Virtual Reality training system.

In the Phase II, in both control and treatment group, the participants were asked to do the Pretest before the training began. After that, they were asked to do the training according to the group. The control group watched the training videos for 30 min, and the treatment group used the VR training system for 30 min, too. Special for the treatment group, the VR training system is also equipped with the special functions for Activity Data Acquisition. That means, during the VR training, the movements of the participants will be recorded in terms of coordinates and hand positions. This activity data log can be then analyzed later. For testing of what will last longer in the trainees' memory, we asked the participants to do the Posttest 3 times, namely, 3 days after the training, 5 days after the training, and 7 days after the training. Then we analyzed the collected data with proper statistical approaches.

In the Phase III, the holistic evaluation framework was also created. In this phase, we analyzed the whole approach and suggested the holistic evaluation frame for industrial training using VR technology. Our suggestion ranges from content creation to evaluation of the hands-on training, including necessary statistical analysis.

## 3.2 *Pretest/Posttest Analysis*

To assess the improvement of the trainees between the treat and control groups, we employed ANCOVA (analysis of covariance), a commonly used statistical method in pretest/posttest designs to analyze the impact of the intervention on dependent variables [12]. The ANCOVA model for this study is:

$$Posttest\ \ scores = a + bGroup + cPretest\ \ scores + e \tag{1}$$

where:

- Posttest scores is the dependent variable
- a is the intercept
- b is the effect size for the independent variable Group (treatment vs. control)
- c is the effect size for the covariate Pretest scores
- e is the error term

By using ANCOVA, we were able to adjust factors for pre-existing group differences by statistically controlling for the effects of the pre-existing variable, which was increasing the accuracy and reliability of the findings.

## 3.3 Activity Data Log

In the assumption that the same industrial training will force the trainees to have the same movement pattern during the training as the trainer, we have to obtain movement data of the trainees. If their movement patterns are alike, this can lead to conclusion that the training was successful. With this concept, we capture movement data of the left hand, right hand, and the head of the trainee. Figure 2 shows the hand and head movements that were being collected within the VR training system.

Thanks to the VR development environment, the Unity [13], it has the Unity VR coaching application that generates a log file in.txt format. The log file is divided into two clusters: the first cluster contains static data information such as IDs and coordinates, while the second cluster contains fresh comments (markups) including data IDs, comment-associated coordinates, and other information.

The Unity software system allows observing in real-time how VR environments are interacting with users. The tracking information of left hand, right hand, and head movements is shown in Table 1. The coordinates X, Y, Z represents the current position of the hands and head in the 3D coordinate system.

In terms of similarity measurement, the Cosine Similarity [14] is used. Its formula between two vectors A and B is defined as:

$$\text{Cosine Similarity (A, B)} = (\text{A} * \text{B})/(||\text{A}|| * ||\text{B}||) \quad (2)$$

where:

- A and B are the two vectors
- A * B is the dot product of the vectors (the sum of the element-wise product of the vectors)
- ||A|| is the magnitude of vector A (the square root of the sum of the squares of the elements of A)

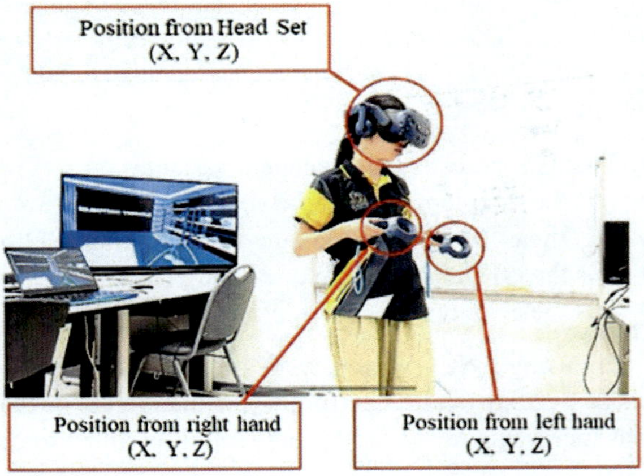


**Fig. 2** Activity Data acquisition by capturing hand and head movements

**Table 1** An excerpt of data from the Unity VR coaching application

| Time | Tracking name | X | Y | Z |
|---|---|---|---|---|
| 00:00 | LeftHand_Controller | 1.1 | 2.6 | 9.1 |
| 00:00 | RightHand_Controlle | 1.4 | 4.1 | 11.5 |
| 00:00 | HMD_Controller | 1.3 | 5.3 | 10.0 |
| 00:01 | LeftHand_Controller | 1.1 | 2.6 | 9.0 |
| 00:01 | RightHand_Controlle | 1.2 | 3.5 | 11.2 |
| 00:01 | HMD_Controller | 1.2 | 5.1 | 10.0 |
| 00:02 | LeftHand_Controller | 1.3 | 2.8 | 11.7 |
| 00:02 | RightHand_Controlle | 1.3 | 3.4 | 8.8 |
| 00:02 | HMD_Controller | 1.2 | 5.0 | 10.4 |

- ||B|| is the magnitude of vector B (the square root of the sum of the squares of the elements of B)

The activity data was obtained as mentioned above. We then combined the movement data into a movement pattern for each trainee. This will be then compare to the standardized movement pattern of a trainer using Cosine Similarity method.

# 4 Results and Discussion

From the conceptual approach described above, we have following results; 1) the interactive VR training system for industrial training, 2) the analysis results of Pretest/Posttest, and 3) the Analysis results of Activity Data. The combination of these results was then evaluated and gave the intended holistic evaluation framework for VR industrial training.

## *4.1 Phase I: Development*

This phase resulted the pedagogical contents and resources under the concept of the ADDIE model. The 3D models were programmed creating 3D realistic simulation for Electrical Vehicle assembly training. In this simulation, the trainees can navigate, reading, watching videos, pick up parts and training to assembly the vehicle parts as if they would do in the real training situation. Figure 3 shows a screenshot of such a VR training system. During the training, the Activity Data was collected and saved for analysis.

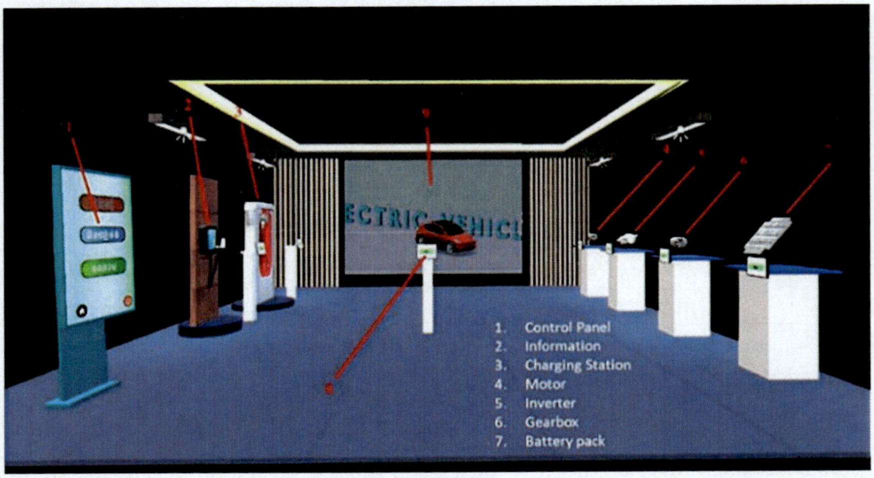


**Fig. 3** VR training system for electrical vehicle assembly

## 4.2 Phase II Evaluation

In this phase, we evaluated the experiments conducted above. The results of the Pretest/Posttest evaluation using ANCOVA method is shown in the Table 2.

**Table 2** Results of the pretest/posttest evaluation using ANCOVA method

| Group | Post | Pre |
|---|---|---|
| Treat | 130 | 90 |
| Treat | 140 | 70 |
| Treat | 150 | 80 |
| Treat | 120 | 30 |
| Treat | 120 | 80 |
| Control | 120 | 30 |
| Control | 90 | 100 |
| Control | 80 | 80 |
| Control | 90 | 20 |
| Control | 110 | 70 |

Ancova: Single Factor

SUMMARY

| *Groups* | *Count* | *Sum* | *Average* | *Variance* |
|---|---|---|---|---|
| Post | 10 | 1150 | 115 | 516.6667 |
| Pre | 10 | 650 | 65 | 783.3333 |

**ANCOVA**

| *Source of Variation* | *SS* | *df* | *MS* | *F* | *P-value* | *F crit* |
|---|---|---|---|---|---|---|
| Between Groups | 12500 | 1 | 12500 | 19.23077 | 0.000357 | 4.413873 |
| Within Groups | 11700 | 18 | 650 | | | |
| Total | 24200 | 19 | | | | |

**Table 3** The Cosine Similarity values between the trainer and the trainees

| Trainee No | Test 1 | Test 2 | Test 3 |
|---|---|---|---|
| T1 | 0.460023 | 0.507265 | 0.760023 |
| T2 | 0.54895 | 0.558739 | 0.77895 |
| T3 | 0.573517 | 0.609064 | 0.803517 |
| T4 | 0.711197 | 0.772936 | 0.811197 |
| T5 | 0.742763 | 0.786582 | 0.842763 |

There are two groups being compared, the "Treat" and the "Control" group. Each group has five participants with two measurements, one taken before treatment ("Pre") and one taken after treatment ("Post").

According to Table 2, the "Between Groups" row shows the sum of squares (SS) and degrees of freedom (df) for the effect of group membership on the outcome variable. In this case, the F-statistic is 19.23077 and the associated p-value is 0.000357, which is less than the conventional threshold of 0.05, indicating that the difference in means between the two groups is statistically significant. The "Total" row shows the total SS and df for the entire analysis.

The left hand, right hand, and head movements' positions were acquired using an activity log file from the VR training system. This data was then calculated with the Cosine Similarity mentioned above. The Cosine Similarity value ranges between $[-1, 1]$, where the most similar ones have the Cosine Similarity value of 1, and vice versa. The Table 3 shows the Cosine Similarity results between the trainer and the trainees.

The Cosine Similarity values of these 5 trainees show in the same direction towards 1. This indicates that the training was getting better. According to the chart represented in Fig. 4, the test number 3 has the highest Cosine Similarity value for all the trainees.

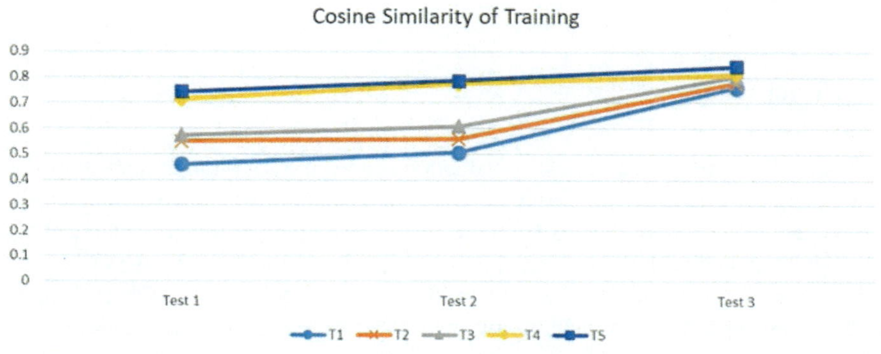


**Fig. 4** Comparison of Cosine Similarity values of the trainer and the trainees

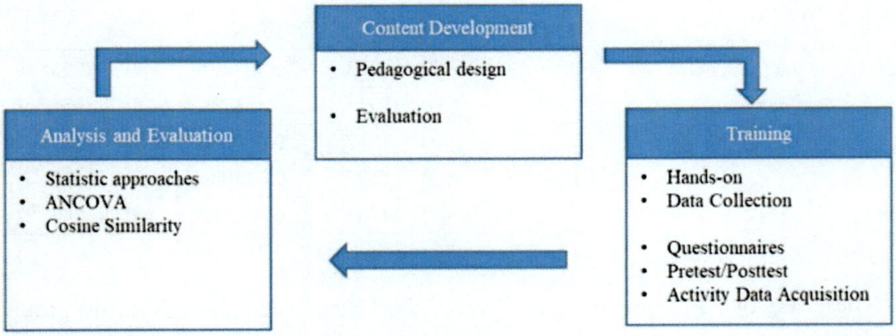


**Fig. 5** The holistic evaluation framework for industrial VR training

### *4.3 Phase III Framework Development*

From the conceptual approach and the results of the experiments, we then constructed the holistic evaluation framework for industrial VR training accordingly. Figure 5 shows our proposed framework.

The development of a VR training system involves creating structured contents of the VR training based on a pedagogic approach and desired learning outcomes. A checklist can be used to ensure that all necessary content is included. In the process of training, several methods can be used for data collecting and evaluation, e.g. Questionnaires, Pretest/Posttest, Activity Data acquisition. Collected data can be then analyzed using suitable statistical methods, e.g. Pretest/Posttest can be analyzed using ANCOVA approach and Activity Data can be then analyzed with Cosine Similarity. The leads to the goal of this framework development that is to create a comprehensive framework for evaluating the VR training program and identifying areas for skill improvement in industrial trainings.

## 5 Conclusion

In this work, we proposed a holistic evaluation framework for industrial VR training and successfully demonstrated with the results of the experiments. In our framework, the process of evaluation begins from the designing of the particular contents. This has to be matched with the current skills and abilities of the trainees, which can be approved by a proper checklist. In the training process, appropriate data collection methods are suggested. For quantitative data, a survey using suitable questionnaires is recommended, whereas the Activity Data acquisition during the VR training is another method for getting the trainees' data. This collected data can be then analyzed in the last process of our framework. According to the data types, the ANCOVA method is recommended for Pretest/Posttest data, and Cosine Similarity method is recommended for the Activity Data. The latter is based on assumption that the industrial training of the same task will force the movements of the trainees' hand and head in the same way. Using this concept, we have to create first a master training pattern of this particular training.

In terms of training effectiveness evaluation of industrial VR training, we will research deeper by using more complex sensors and analytic methods. Other sensors such as vital signs monitoring systems or eye gazing systems are a good complement to our research, as well as movement and posture analysis of training. In addition, gaining more participants will also help us to improve accuracy of the training effectiveness measurement.

**Acknowledgements** This work is done at Digital Media and Interaction Research Laboratory (DMI), Faculty of Informatics, Burapha University. We are deeply grateful for the grant and supports from the Faculty of Informatics. Furthermore, we want to thank the Erasmus+ Project "Education Training for Automation 4.0 in Thailand" (ETAT) No.610154-EPP-1-2019-1-DE-EPPKA2-CBHE-JP for networking and collaboration support.

## References

1. Denny EB (1991) Connectionism and the specter of representationalism, In: Horgan TE, Tienson JL (eds) Connectionism and the philosophy of mind, pp 417–436. Kluwer Academic Publishers
2. Shiva P, et al (2017) Evaluating 360-virtual reality for mining industry's safety training, In: HCI International 2017 – Posters' Extended Abstracts, pp 555–561
3. Matthew JD et al (2021) Toward a framework for developing virtual reality skills training in human services. J Technol Hum Serv 39(3):295–313
4. Boonmee P, Jantarakongkul B, Jitngernmadan P (2020) VR training environment for electrical vehicle assembly training in EEC. In: 2020 - 5th International conference on information technology (InCIT)
5. Radianti J, et al (2020) A systematic review of immersive virtual reality applications for higher education: design elements, lessons learned, and research agenda. Comput Educ 147
6. Molenda M (2003) In search of the elusive ADDIE model. Perform Improv 42:34–36
7. Smidt A et al (2009) The Kirkpatrick model: a useful tool for evaluating training outcomes. J Intellect Dev Disabil 34(3):266–274
8. Choi KS (2019) Virtual reality wound care training for clinical nursing education: an initial user study. In: 2019 IEEE conference on virtual reality and 3D user interfaces (VR)
9. Concannon D, Flynn R, Murray N (2019) A quality of experience evaluation system and research challenges for networked virtual reality-based teleoperation applications, In: Proceedings of the 11th ACM workshop on immersive mixed and virtual environment systems, pp 10–12. Association for Computing Machinery, Amherst
10. Zhang Y, et al (2020) Virtual reality applications for the built environment: research trends and opportunities. Autom Constr **118**
11. Hou G, Dong H, Yang Y (2017) Developing a virtual reality game user experience test method based on EEG signals. In: 2017 5th international conference on enterprise systems (ES), pp 227–231

12. Keselman HJ et al (1998) Statistical practices of educational researchers: an analysis of their ANOVA, MANOVA, and ANCOVA analyses. Rev Educ Res 68(3):350–386
13. Wikipedia: Unity (game engine). https://en.wikipedia.org/w/index.php?title=Unity_(game_engine)&oldid=1132951579. Accessed 16 Jan 2023
14. Cosine similarity. https://en.wikipedia.org/wiki/Cosine_similarity. Accessed 16 Jan 2023

# Rice Diseases Recognition Using Transfer Learning from Pre-trained CNN Model

**Wittawat Hamhongsa, Rungrat Wiangsripanawan, and Pairat Thorncharoensri**

**Abstract** This research aims at applying the well-known pre-trained convolutional neural network (CNN) image classification algorithms such as InceptionV3, Xception, ResNetV2, InceptionResNetV2, and DenseNet to classify the five diseases of rice's leaves in Thailand. Data sets used are from 3 sources: UCI database, Rice Leaf Disease Image Samples dataset and images collected by authors in Thailand between 2018–2020. Our initial experimental result shows that using the pre-trained CNN models to classify the rice leaves disease seems to be possible with a greater number of images required. Therefore, the image data augmentation technique is used to add the number of images into the dataset. The experimental result with data augmentation shows that it could increase rice disease classification efficiency up to 16.533% (especially for the ResNet50V2 model).

**Keywords** Convolutional neural network · Deep learning · Pre-trained CNN model · Rice diseases · Transfer learning

## 1 Introduction

As one of the top three rice exporters globally [1], Thailand is one of the highest rice consumption countries [2]. Hence, rice has become an industrial crop for Thai people. Increasing rice yield is challenging because there are several factors that result in its productivity. Eliminating rice disease on time is one of the main factors. However, this requires experts or experienced farmers to identify diseases from the rice leaves. In agriculture, convolutional neural networks techniques (CNNs), the best technique for computer vision in the deep learning area, have been used to solve

W. Hamhongsa · R. Wiangsripanawan (✉)
Department of Computer Science, School of Science, King Mongkut's Institute of Technology Ladkrabang, Bangkok, Thailand
e-mail: rungrat.wi@kmitl.ac.th

P. Thorncharoensri
On Technology Pty. Ltd., Perth, Australia
e-mail: pthorncharoensri@codisto.com

 
P. Meesad et al. (eds.), *Proceedings of the 19th International Conference on Computing and Information Technology (IC2IT 2023)*, Lecture Notes in Networks and Systems 679,
https://doi.org/10.1007/978-3-031-30474-3_16

several problems using images such as identifying the herbarium specimens [3], distinguishing a healthy plant from a diseased plant [4–6] or identifying and assessing the severity of biological stress on coffee leaves [7]. For rice plants, Shrivastava et al., [8] proposed a method to classify rice diseases using transfer learning with AlexNet model with SVM. Chen et al. [9] studied the detection of rice diseases based on deep transfer learning using a public image database and the dataset from real surroundings.

A dataset is important to the CNN model but there are not many rice disease image datasets that are publicly available. The existing ones, their datasets are either small or consist of merely noncomplex background images or images captured in the laboratory, not from the real rice field. Hence, this research also constructs another dataset from photos captured from the rice fields in Nakhon Pathom, a province in central Thailand. Still, the number of images seemed to be too small.

The image data augmentation technique [5, 10] is another interesting technique that could increase the number of images. For example, Taylor et al. [11] used it to increase the number of airplane images used in their classification and Zhang et al. [12] enhanced the image of the leaves with an image processing process showing that data augmentation technique can enhance the model's perception.

There are three contributions presented in this paper. First, a new dataset for healthy and rice leaf disease images found from real rice fields in Thailand is created. Second, the experimental result shows the likelihood of using the nine well-known pre-trained CNN models to distinguish the five rice leaf diseases commonly found in Thailand with a greater number of dataset images. Finally, the image data augmentation technique is applied in the pre-processing to increase the number of images for better accuracy of all models.

## 2 Background Knowledge

### *2.1 InceptionV3*

InceptionV3 [13] is a convolutional network used for object detection and image classification developed from GoogLeNet [14]. InceptionV3 architecture scales up from Inception architecture. InceptionV3 consists of 159 depths and 23,851,784 parameters. There are symmetric and asymmetric building blocks, convolutions, max pooling, average pooling, dropouts, concats and fully connected layers. With extensively use of Batchnorm in the model and application to activate inputs, the loss is calculated via Softmax. Examples of using InceptionV3 are flower classification [15], fresh tea leaves classification [16], and soil health analysis and fertilizer prediction [17].

## 2.2 Xception

Xception [18] has been developed from Inception architecture. Its name came from "Extreme Inception". The idea is to replace the Inception modules with a depth-wise separatable convolution and a maximal number of towers. Xception consists of 22,910,480 parameters and 126 depths. Xception is a little more effective than InceptionV3 on a larger image classification dataset comprising 350 million images and 17,000 classes. Similar to the InceptionV3, the model is made up of symmetric and asymmetric building blocks. An example of using InceptionV3 is a segmentation of abnormal leaves of hydroponic lettuce based on DeepLabV3 + for robotic sorting by Wu et al. [19].

## 2.3 ResNetV2

ResNetV2 [20] has been developed from ResNet [21] which its prominent point is ResNet uses batch normalization at its core. Batch normalization adjusts the input layer, resulting in an increase in the network's performance and resolving the problem of covariate shift. ResNet protects the network from vanishing gradient problems using identity connection. ResNet using a bottleneck residual block design results in increasing performance of the network. The differences between ResNetV2 and ResNet are ResNetV2 uses batch normalization and ReLU activation to the input layer then multiplies with the weight matrix (convolution operation). Examples of using ResNetV2 are apple tree leaf diseases classification by Alsayed et al. [22] and paddy leaf disease detection by Islam et al. [23]. The numbers 50, 101, and 152 in ResNet50V2, ResNet101V2, and ResNet152V2 indicate the convolutions layer used in each RestNetV2 model.

## 2.4 InceptionResNetV2

InceptionResNetV2 [24] is a combination of the InceptionV3 and the ResNet models. The architecture of the network is considerably deeper than the previous InceptionV3. At the Inception-Resnet block, the residual connection is combined with multiple sizes of convolutional filters. Using the residual connection is not only to avoid the degradation problem caused by deep structures but also to promote reduced training time. Example works using InceptionResNetV2 are classifying plant leaf disease by Naveenkumar [25] and Hassan et al. [26], and paddy leaf disease detection by Islam et al. [23].

## 2.5 DenseNet

The distinctive point of DenseNet [27] is densely connected architecture. Each convolutional layer has connected directly except the input layer. Dense Blocks have a characteristic of feed-forward, that is the output from the previous convolutional layer will be sent to the next layer. DenseNet has interesting advantages such as strengthening feature propagation, reducing vanishing gradient problems, supporting feature reuse, and greatly lessening the number of parameters. Examples of using DenseNet are leaf disease identification by Verma et al. [28] and plant leaf classification by Park et al. [29] and by Swaminathan et al. [30]. The numbers 121, 161, and 201 in DenseNet121, DenseNet169, and DenseNet201 models indicate the convolutions layer used in each DenseNet model.

## 2.6 Transfer Learning

The benefit of transfer learning is to shorten the duration and resources in training. A pre-trained model with transfer learning is more precise than sketch training which trains a model with small data in the first place [31]. Transfer learning [32] is the weight transfer of CNN models learned from a big dataset that most of the models had been trained by ImageNet datasets. The ImageNet dataset spans 1,000 object classes and contains 1,281,167 training images, 50,000 validation images, and 100,000 test images [33].

## 2.7 Image Data Augmentation Technique

Deep learning neural network performance often improves with the amount of data available. The image data augmentation techniques [5, 10] increase the image dataset by randomly generating a new image from the original image. The image data augmentation techniques are primarily used during image processing. A new image is created from the original image by performing some operations such as rotation, zooming, flipping, and other process involving brightness and color filtering. The object seen by the model and the real object must be consistent, for example, a horizontal flip of a dog picture might make sense. Because it may be taken from the left or right. A vertical flip of a dog picture does not make sense and may not be appropriate since the model is not very likely to see the dog upside down. Therefore, when choosing an Image data augmentation technique, consideration should be given to the nature of the image and the loss of important image features that will degrade the model's performance. The CNN model can learn features that are invariant to their location in the image. The image data augmentation technique can additionally support this change in the invariant way to deal with learning. Moreover, it helps the model to learn the highlights that are likewise invariant to changes. For example, sequential direction image, light levels in image, and more.

## 3 Experiment

Five rice left diseases that were frequently found in Thailand [34] and our rice fields during the experimental period were chosen for classification. They are Blast disease (Fig. 1 (a)), Brown Spot disease (Fig. 1(b)), Bacterial Leaf Blight disease (Fig. 1 (c)), Rice Tungro disease (Fig. 1 (d)), and Narrow Brown Spot disease (Fig. 1 (e)). There is also a healthy class. Totally, there are six classes: one healthy and five disease classes.

Our experiment consists of 3 main tasks. The first task is dataset preparation. Our original dataset was prepared from 3 sources: the UCI database [35], the Rice Leaf Disease Image Samples dataset [36], and our own dataset which was collected from the real rice fields in 2018–2020. The second task is to conduct an experiment (trained

**Fig. 1** Rice diseases (a) Blast disease (b) Brown Spot disease (c) Bacterial Leaf Blight disease (d) Rice Tungro disease (e) Narrow Brown Spot disease

and tested) using the original dataset with the pre-trained models available in Keras. The final task is to employ image augmentation techniques in data preprocessing to improve the classification performance of all chosen models. Two datasets (DatasetA and DatasetB) were generated using different data augmentation techniques.

## *3.1 Dataset Preparation*

As above mentioned, there are 3 sources of the dataset. The first dataset (Dataset 1) is from the UCI database [35] which consists of white background images captured under laboratory conditions. The second dataset (Dataset 2) is from the Rice Leaf Disease Image Samples dataset [36]. It consists of complex background images captured under field conditions. The last dataset is our collection of images from the rice fields in Nakhon Pathom, a province in Central Thailand.

### 3.1.1 Dataset Construction form the Rice Fields

Dataset 1 and Dataset 2 are from the public database that does not cover some rice leaf diseases found in Thailand. Also, by the time our experiment was conducted, the databases for Thai rice leaf diseases were rare and not publicly available. Therefore, the first goal of our research is to construct a dataset from real rice fields to increase the number of images. The rice fields in Banglen District, Nakhon Pathom were chosen. First, 2,712 photos of suspected either healthy or sick leaves were taken by smartphones from the rice fields in 2018–2020. Second, a suspicious area was cropped from each photo. Third, each cropped image was initially classified into each category by the author. Note that all images were complex and had discontinuous luminous intensity since they were captured under real agriculture surrounding and there is no other image processing apart from cropping at the stage yet.

All classified 2,712 images were sent to the agricultural research officer from Rice Department, the Ministry of Agriculture and Cooperative, to verify their correctness. Note that the research officer's task was to examine and confirm our classification using her expertise. The wrongly classified images were removed and sent back to be re-classified since the scars on the leaves look similar for a few diseases. The images that passed the second classification from the researcher were sent again to the research officer. If the result was still wrong, this image would be removed from the dataset.

After a long process of data collection and classification, 1,888 out of 2,712 were approved by the agricultural scholar.

**Table 1** Number of images in dataset

| Diseases | Dataset 1 | Dataset 2 | Our Dataset | Total |
|---|---|---|---|---|
| Healthy | – | – | 899 | **899** |
| Blast disease | – | 1440 | 315 | **1,755** |
| Brown Spot disease | 40 | 1600 | 240 | **1,880** |
| Bacterial Leaf Blight disease | 40 | 1584 | 155 | **1,779** |
| Rice Tungro disease | – | 1308 | – | **1,308** |
| Narrow Brown Spot disease | – | – | 279 | **279** |
| Capture conditions | Laboratory | Field | Field | - |

#### 3.1.2 The Original Dataset

Dataset 1, Dataset 2, and our dataset are combined and called the *original dataset*. Table 1 shows the number of images of each class in each dataset including their capture conditions. It is noted that all images were also resized to 224 × 224 pixels.

### *3.2 CNN Model and Their Setting*

The Pre-trained CNN models used in our experiment are Xception, ResNet50V2, ResNet101V2, ResNet152V2, InceptionV3, InceptionResNetV2, DenseNet121, DenseNet169 and DenseNet201. See Table 2 for the models' size, parameters, and depth. The transfer learning weights are from the ImageNet. The batch size, the epoch, and the learning rate are set to 64, 20 and 0.0001 respectively. The dataset images were split randomly into train and test sets with a ratio of 80:20. The performance of each model is measured using the accuracy value. Each model was run 10 times to calculate the average accuracy value. It is noted that the train and test sets were chosen randomly each time.

### *3.3 Experimental Result of the Original Dataset*

Table 3 shows the accuracy value of each model with the original dataset. The values from all models are between 0.4977 – 0.605 which is not impressive. More work is needed to be done to improve the performance. The lowest and the highest performance are DenseNet121 and DenseNet201. Other models, apart from the DenseNet family, their values are not much different (0.5312 to 0.5543).

**Table 2** Pre-trained CNN model

| Model | Size | Parameters | Depth |
|---|---|---|---|
| Xception | 88 MB | 22,910,480 | 126 |
| ResNet50V2 | 98 MB | 25,613,800 | - |
| ResNet101V2 | 171 MB | 44,675,560 | - |
| ResNet152V2 | 232 MB | 60,380,648 | - |
| InceptionV3 | 92 MB | 23,851,784 | 159 |
| InceptionResNetV2 | 215 MB | 55,873,736 | 572 |
| DenseNet121 | 33 MB | 8,062,504 | 121 |
| DenseNet169 | 57 MB | 14,307,880 | 169 |
| DenseNet201 | 80 MB | 20,242,984 | 201 |

**Table 3** Accuracy of original dataset

| Model | Accuracy | Model | Accuracy |
|---|---|---|---|
| Xception | 0.54570 | InceptionResNetV2 | 0.54297 |
| ResNet50V2 | 0.55156 | DenseNet121 | 0.49766 |
| ResNet101V2 | 0.55430 | DenseNet169 | 0.52219 |
| ResNet152V2 | 0.53125 | DenseNet201 | 0.60586 |
| InceptionV3 | 0.54766 | | |

### *3.4 Performance Improvement*

To maximize the effectiveness of CNN learning, more images are required. Instead of obtaining more images from the rice field, which was difficult during the Covid situation and would take much longer time to process. Basic image data augmentation techniques in Keras (imageDataGenerator) was used. Two different methods are used to create two new groups of datasets (Dataset A and Dataset B).

In the first method, basic augmentation techniques which are rotation, height shifting, width shifting, zooming, horizontal flipping, and vertical flipping are used with a small random range of parameter values. As shown in Fig. 2, rather than doing one augmentation process (A1) to generate one new dataset (DatasetAx2), the process is run nine times to generate nine sets of images (A1 to A9) to construct nine datasets (DatasetAx2 – DatasetAx10). Note that as imageDataGenerator in Keras randomly chooses augmentation parameters including their values, the augmented image may be similar to the original image.

In the second method, parameters related to the light and color of the images are added to the imageDataGenerator. Unlike the first method which the same parameters are used, each expansion uses three different parameters with bigger random range values to generate different sets of images (B1 to B9) to construct nine datasets (DatasetBx2 to DatasetBx10). See Fig. 3 for more details.

These eighteen datasets (DatasetAx2 – DatasetAx10 and DatasetBx2 – DatasetBx10 were experimented with all pre-trained CNN models under the same

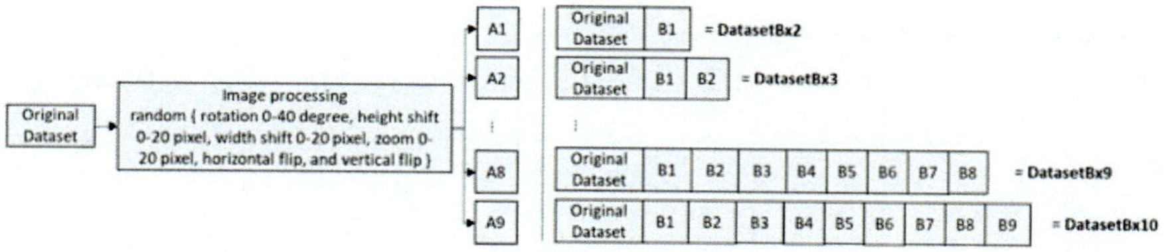


**Fig. 2** Image data augmentation technique for DatasetA

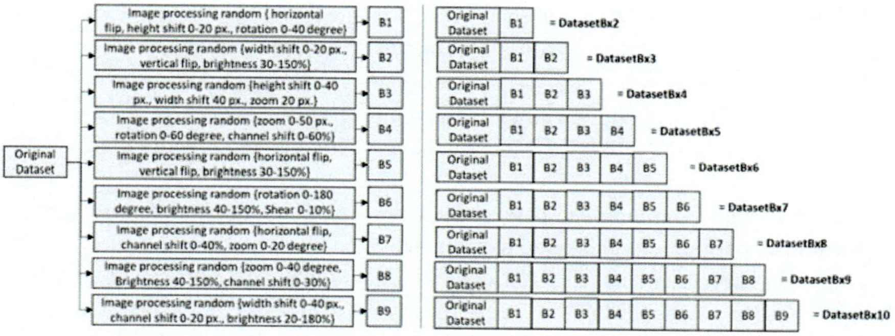


**Fig. 3** Image data augmentation technique for DatasetB

settings as in Subsection 3.2 The main experimental goal is to study the improvement of each model's accuracy after applying basic image augmentation methods to our original dataset. The subsequence goals are to study whether more numbers of augmentation images will always continuingly increase the accuracy performance and to study whether different usage of imageDataGenerator affects the performance.

## 4 Results and Discussion

Table 4 shows the accuracy of each model with different datasets: the original dataset and the augmented datasets (Dataset A and Dataset B). For the original dataset, DenseNet201 seems to be the best, its accuracy value although low (0.60586) outperforms other models by 5 to 10 percent whereas Densenet121 gives the lowest value (0.49766). Other models' accuracy values are sorted ascendingly as follows: ResNet101V2 (0.55430), ResNet50V2 (0.55156), InceptionV3 (0.54766), Xception (0.54570), InceptionResNetV2 (0.54297), ResNet152V2 (0.53125) and DenseNet169 (0.52219).

For the augmented Dataset A, the accuracy value for each model is the maximum accuracy value when applying 80:20 training-and-testing ratio to its nine datasets (DatasetAx2 to DatasetAx10). A similar approach is done with the augmented

Table 4 Maximum Accuracy value of each dataset

| Model | Original Dataset | DatasetA | DatasetB | Increase from Original Dataset % |
|---|---|---|---|---|
| Xception | 0.54570 | 0.68867 | **0.70183** | 15.613 |
| ResNet50V2 | 0.55156 | 0.69102 | **0.71689** | **16.533** |
| ResNet101V2 | 0.55430 | 0.69766 | **0.70338** | 14.908 |
| ResNet152V2 | 0.53125 | **0.68398** | 0.68183 | 15.273 |
| InceptionV3 | 0.54766 | **0.68984** | 0.67305 | 14.219 |
| InceptionResNetV2 | 0.54297 | **0.69450** | 0.67070 | 15.153 |
| DenseNet121 | 0.49766 | 0.62914 | **0.62944** | 13.178 |
| DenseNet169 | 0.52219 | 0.66938 | **0.67273** | 15.055 |
| DenseNet201 | 0.60586 | 0.68083 | **0.69089** | 8.503 |

Dataset B. The Maximum value is chosen because it is derived from the best augmentation parameter values and dataset size.

With the augmented datasets, DenseNet201 is no longer the best model on either dataset. Besides, DenseNet121 still keeps the lowest results: 0.62914 (Dataset A) and 0.62944 (Dataset B) so DenseNet121 seems to be not a good candidate in our classification. For Dataset A, the performance of models excluding DenseNet121 is listed as follows: ResNet101V2 (0.69766), InceptionResNetV2 (0.69450), ResNet50V2 (0.69102), InceptionV3 (0.68984), Xception (0.68867), DenseNet201 (0.68083) and DenseNet169 (0.66938). Hence, the augmented methods employed in Dataset A resulted in 67 to 70 percent accuracy. The performance of each model in Dataset B is as follows: ResNet50V2 (0.71689), ResNet101V2 (0.70338), Xception (0.70183), DenseNet201 (0.69089), ResNet152V2 (0.68183), InceptionV3 (0.67305), DenseNet169 (0.67273) and InceptionResNetV2 (0.67070), and DenseNet121 (0.62914). Hence, excluding DenseNet121, the models' accuracy value is ranged from 67 to 72 percent. ResNet50V2 and ResNet101V2 performs well in both augmented datasets.

The accuracy values of all models are increased between 8.503 to 16.533 percent in both Dataset A and Dataset B. The highest increment occurs with ResNet50V2 (16.533) followed by Xception (15.613) and ResNet152V2 (15.273) respectively. Whereas the lowest increment is from DenseNet201 (8.503), DenseNet201 provides the best performance in the original dataset but does not do well with the augmented datasets. One assumption is that some augmentation attributes in the datasets are already learned since the original dataset whereas other models have not learned this knowledge yet. According to the incremental values from our experiment, our basic augmentation techniques used can increase the accuracy values by 13 to 17 percent for models apart from DenseNet201.

As seen from the graphs in Fig. 4, the accuracy values in Dataset A are increased significantly since the first augmentation (×2). To illustrate, the accuracy values of all ResNet models, Xception, and Inception are increased by around 10 percent since the first augmentation. After that, the values still increase but at a much lower rate

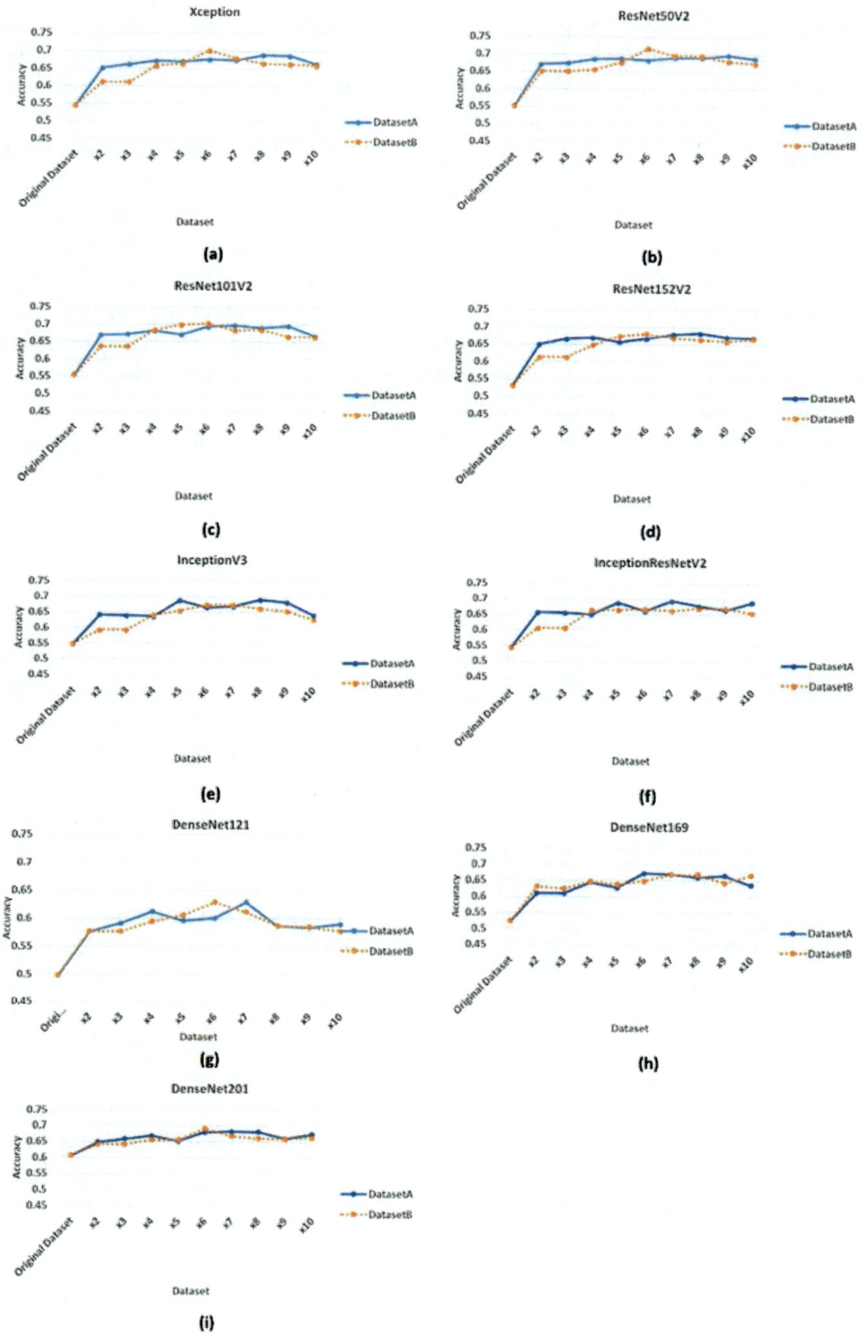


**Fig. 4** Graphs showing comparison of the accuracy value of each model for the original dataset, Dataset A and Dataset B

(less than 5 percent). The peak values mostly are at either $\times$ 5 to $\times$ 7. The values are dropped and then steady or decreased. Unlike Dataset A, the accuracy values for the models in Dataset B gradually increased with the bigger dataset size. They seem to finally reach their steady points as well.

To summarize, (1) DenseNet201 is the best model if only the original dataset is used. (2) DenseNet168 is the least efficient model used in any of our datasets. (3) Basic data augmentation techniques can boost the efficiency of all models. The highest is around 70 percent. Under our both augmented techniques, ResNet50V2, ResNet101V2, and Xception are quite attractive for further study such as using them as models to fine-tune. (4) Accuracy values depend on how data augmentation techniques are deployed. The more complex augmented technique creates more complex images which result in a bigger usable dataset size that delivers higher accuracy. (5) DenseNet201 is not suitable with the basic augmentation techniques such as the ones used in our augmented datasets. A more complex data augmentation approach should be deployed with this model.

## 5 Conclusion

In this work, a new dataset consisting of 1,888 images of healthy and four rice leaf diseased is constructed from photos taken in real rice fields in Thailand during 2018–2020. Images in this dataset are verified by an expert from the Rice department. It then combines with two public datasets and is used as the experimental dataset for the nine pre-trained CNN models using transfer learning to classify the six classes of rice leaf: one healthy class and five disease classes. Accuracy value is used as the performance metric. The first experiment gives a range of accuracy between 49.766 to 60.586 percent. Hence, two data augmentation methods are employed to increase the number of images. The first method uses the position transformation randomly chosen by Keras to create more images (Dataset A). The second method uses both position-transformation and color-transformation techniques which are customarily chosen (Dataset B). For each augmentation method, the experiment is done with nine different datasets. The experimental result shows that data augmentation can improve the accuracy of all models. The new accuracy values are between 62.914 to 71.689 percent. It is not necessary that models that provide higher accuracy in the original dataset will result in higher accuracy in the augmented datasets (DenseNet201). More complex data augmentation techniques and a bigger dataset size tend to offer higher accuracy values. The highest incremental accuracy is from ResNet50V2 with 16.533 percent.

## 6 Future Work

Future work can be done in a variety of ways. The first one is to construct more images from the real rice fields or combine them with more publicly available datasets since the image data augmentation techniques are only the simulation of the data. It is interesting to fine-tune this dataset with the Densenet101 model since it gives the highest performance with the original dataset. The second option is to use the basic data augmentation method and fine-tune the pre-trained CNN models so that more specific features of rice leaves are learned. The proposed models are ResNet50V2, ResNet101V2, or Xception. The third extension is to apply more advanced data augmentation techniques such as Generative adversarial networks (GANs) to pre-process data before sending it to models such as Densenet101, ResNet50V2, ResNet101V2, and Xception. In addition, the dataset may include more photos that show traces of pests such as insects and worms, etc. Each task is challenging in different ways.

**Acknowledgements.** We thank K. Wanporn Khemmuk, the director of Rice Protection Science Group, Division of Rice Research and Development, Rice Department of Thailand for her kindness to assist us in examining and verifying our classified images collected from the rice fields. Without her tremendous help, we would not have had our dataset.

## References

1. Muthayya S et al (2014) An overview of global rice production, supply, trade, and consumption. Ann. N Y Acad. Sci. 1324(1):7–14
2. Berno T, Dentice G, Wisansing JJ (2019) Kin kao laew reu young ('have you eaten rice yet')?: A new perspective on food and tourism in Thailand. In: Park E, Kim S, Yeoman Ian (eds) Food tourism in Asia. Springer Nature Singapore, Singapore, pp 17–30. https://doi.org/10.1007/978-981-13-3624-9_2
3. Carranza-Rojas J et al (2017) Going deeper in the automated identification of Herbarium specimens. BMC Evol Biol 17(1):1–14
4. Mohanty SP, Hughes DP, Salathé M (2016) Using deep learning for image-based plant disease detection. Front. Plant Sci. 7:1419
5. Sladojevic S, Arsenovic M, Anderla A, Culibrk D, Stefanovic D (2016) Deep neural networks based recognition of plant diseases by leaf image classification. Comput. Int. Neuroscie 2016:1–11. https://doi.org/10.1155/2016/3289801
6. Ferentinos KP (2018) Deep learning models for plant disease detection and diagnosis. Comput. Electron. Agric. 145:311–318
7. Esgario JG, Krohling RA, Ventura JA (2020) Deep learning for classification and severity estimation of coffee leaf biotic stress. Comput Electron Agric 169:105162
8. Shrivastava, V.K., et al.: Rice plant disease classification using transfer learning of deep convolution neural network. In: International Archives of the Photogrammetry, Remote Sensing & Spatial Information Sciences (2019)
9. Chen J et al (2020) Detection of rice plant diseases based on deep transfer learning. J. Sci. Food Agric. 100(7):3246–3256
10. Alzubaidi L et al (2021) Review of deep learning: Concepts, CNN architectures, challenges, applications, future directions. J. big Data 8(1):1–74

11. Taylor, L., Nitschke, G.: Improving deep learning with generic data augmentation. In: 2018 IEEE Symposium Series on Computational Intelligence (SSCI) 2018, pp. 1542–1547 (2018)
12. Zhang, C., et al.: A convolutional neural network for leaves recognition using data augmentation. In: 2015 IEEE International Conference on Computer and Information Technology; Ubiquitous Computing and Communications; Dependable, Autonomic and Secure Computing; Pervasive Intelligence and Computing 2015, pp. 2143–2150. (2015)
13. Szegedy, C., et al.: Rethinking the inception architecture for computer vision. In: Proceedings of the IEEE Conference on Computer Vision and Pattern Recognition, pp. 2818–2826. (2016)
14. Szegedy, C., et al.: Going deeper with convolutions. In: Proceedings of the IEEE Conference on Computer Vision and Pattern Recognition 2015, pp. 1–9 (2015)
15. Xia, X., Xu, C., Nan, B.: Inception-v3 for flower classification. In: 2017 2nd international conference on image, vision and computing (ICIVC) 2017, pp. 783–787 (2017)
16. Qian, Y., et al.: Fresh tea leaves classification using inception-V3. In: 2019 IEEE 2nd International Conference on Information Communication and Signal Processing (ICICSP) 2019, pp. 415–419 (2019)
17. Meenakshi, M., Naresh, R.: Soil health analysis and fertilizer prediction for crop image identification by Inception-V3 and random forest. Remote Sensing Applications: Society and Environment, 100846 (2022)
18. Chollet, F.: Xception: Deep learning with depthwise separable convolutions. In: Proceedings of the IEEE Conference on Computer Vision and Pattern Recognition 2017, pp. 1251–1258. (2017)
19. Wu Z et al (2021) Segmentation of abnormal leaves of hydroponic lettuce based on DeepLabV3+ for robotic sorting. Comput. Electron. Agric. 190:106443
20. He K, Zhang X, Ren S, Sun J (2016) Identity mappings in deep residual networks. In: Leibe B, Matas J, Sebe N, Welling M (eds) Computer Vision – ECCV 2016: 14th European Conference, Amsterdam, The Netherlands, October 11–14, 2016, Proceedings, Part IV. Springer International Publishing, Cham, pp 630–645. https://doi.org/10.1007/978-3-319-46493-0_38
21. He, K., et al.: Deep residual learning for image recognition. In: Proceedings of the IEEE Conference on Computer Vision and Pattern Recognition 2016, pp. 770–778 (2016).
22. Alsayed A, Alsabei A, Arif M (2021) Classification of apple tree leaves diseases using deep learning methods. Int. J. Comput. Sci. Netw. Secur. 21(7):324–330
23. AshiqulIslam Md, Nymur Md, Shamsojjaman M, ShazidHasan Md, Shahadat TK (2021) An automated convolutional neural network based approach for paddy leaf disease detection. Int. J. Adv. Comput. Sci. Appl. 12(1):280–288. https://doi.org/10.14569/IJACSA.2021.0120134
24. Szegedy C, Ioffe S, Vanhoucke V, Alemi A (2017) Inception-v4, inception-resnet and the impact of residual connections on learning. Proc. AAAI Conf. Artif. Intel. 31(1):4278–4284. https://doi.org/10.1609/aaai.v31i1.11231
25. Naveenkumar, M., et al.: InceptionResNetV2 for plant leaf disease classification. In: 2021 Fifth International Conference on I-SMAC (IoT in Social, Mobile, Analytics and Cloud)(I-SMAC) 2021, pp. 1161–1167 (2021)
26. Hassan SM et al (2021) Identification of plant-leaf diseases using CNN and transfer-learning approach. Electronics 10(12):1388
27. Huang, G., et al.: Densely connected convolutional networks. In: Proceedings of the IEEE Conference on Computer Vision and Pattern Recognition 2017, pp. 4700–4708 (2017).
28. Verma R, Singh V (2022) Leaf disease identification using DenseNet. In: AmitaDev SS, Agrawal AS (eds) Artificial Intelligence and Speech Technology: Third International Conference, AIST 2021, Delhi, India, November 12–13, 2021, Revised Selected Papers. Springer International Publishing, Cham, pp 500–511. https://doi.org/10.1007/978-3-030-95711-7_42
29. Park YM et al (2018) Classification method of plant leaf using densenet. J. Korea Multimed. Soc. 21(5):571–582
30. Swaminathan A, Varun C, Kalaivani S (2021) Multiple plant leaf disease classification using densenet-121 architecture. Int. J. Elect. Eng. Technol. (IJEET) 12(5):38–57
31. Bengio, Y.: Deep learning of representations for unsupervised and transfer learning. In: Proceedings of ICML Workshop on Unsupervised and Transfer Learning 2012, pp. 17–36 (2012)

32. Tan, C., et al.: A survey on deep transfer learning. In: International Conference on Artificial Neural Networks 2018, pp. 270–279 (2018)
33. Deng, J., et al.: Imagenet: a large-scale hierarchical image database. In: 2009 IEEE Conference on Computer Vision and Pattern Recognition 2009, pp. 248–255 (2009)
34. Rice Department of Thailand, Rice Knowledge Bank. https://www.ricethailand.go.th/rkb3. Accessed 13 Feb 2023
35. Newman, D.J.: UCI repository of machine learning database. http://www.ics.uci.edu/~mlearn/MLRespository.html (1998)
36. Sethy, P.K.: Rice Leaf Disease Image Samples Mendeley Data (2020)https://doi.org/10.17632/fwcj7stb8r.1

# Machine Learning-Based Methods for Identifying Bug Severity Level from Bug Reports

**Kamthon Sarawan, Jantima Polpinij, and Bancha Luaphol**

**Abstract** Bug reports constitute an important source of information that may be utilized to remedy bugs. One aspect of bug report analysis is determining the severity level of the reported problem. Nevertheless, manually assigning severity to a large number of bug reports is time-consuming, and human analysis of the severity of software bugs may be subject to biased. Consequently, automated analysis is required for detecting the severity of software defects concealed inside bug reports. Even though there have been several attempts to tackle the problem and a number of machine learning and text mining approaches have been applied, the performance of the solutions may vary depending on the datasets employed. As a result, this is the objective of our research, in which we intend to evaluate the efficacy of many machine learning approaches (i.e. Logistic Regression, Random Forest, Support Vector Machine, and Long Short-Term Memory) for severity analysis on multiple bug report datasets (i.e. Core, Firefox, Thunderbird, and Bugzilla) from the Mozilla bug tracking system. It should be emphasized that our research is based on a binary classification scheme in which bug reports are classified into two classes: severe class and non-severe class. The experimental results revealed that the LR algorithm performed well in analyzing the severity of bug reports in the Core, Firefox, and Bugzilla datasets, whereas the LSTM approach performed well for the Thunderbird dataset.

**Keywords** Bug report · Bug severity analysis · Machine learning algorithm

K. Sarawan · J. Polpinij (✉)
Faculty of Informatics, Mahasarakham University, Mahasarakham 44150, Thailand
e-mail: jantima.p@msu.ac.th

K. Sarawan
e-mail: 65011294001@msu.ac.th

B. Luaphol
Faculty of Administrative Science, Kalasin University, Kalasin, Thailand
e-mail: bancha.lu@ksu.ac.th



P. Meesad et al. (eds.), *Proceedings of the 19th International Conference on Computing and Information Technology (IC2IT 2023)*, Lecture Notes in Networks and Systems 679,
https://doi.org/10.1007/978-3-031-30474-3_17

## 1 Background

Despite being tested and having problems corrected before going live, all software still contains some undiscovered faults or defects. Finding software defects is often the responsibility of the software testing or software development teams when the software is small. However, identifying defects in big software needs end-users to submit software issues that they have discovered. End-users of software may be found all around the world. As a result, gathering software defects from them may never be straightforward. A bug tracking system (BTS) has been built and provided in response to this issue [1–4]. The purpose of establishing this system is to use it as a better method for gathering software bug reports from end users all around the world.

After the bug reports are delivered to the BTS system, a "*bug triager*" examines them to verify their validity, correctness, significance, severity, and duplication. This is known as the "bug triage" process and determining bug severity is an important task in the bug triage process [3, 4]. Later, the bug reports are forwarded to the appropriate developers in order to resolve software issues. It is well-known that a bug triager is a software specialist who investigates and refines reported software issues.

Bug severity is an essential indicator that may be used to identify issues that require quick attention. The severity of a bug is defined as the impact of the problem on program functioning [5]. However, manually assigning severity to a large number of bug reports is a time-consuming and tiresome task. Furthermore, categorization accuracy is dependent on the knowledge and expertise of the bug triager who assessed the problem, and analyzing the severity of software defects by humans may be biased. As a result, determining the severity of software bugs reported to a BTS system necessitates the use of automated analysis. The first effort to automate the bug triaging process was found in [6]. They recommended using a supervised machine learning (ML) algorithm to determine which developer should be assigned a bug. Menzies et al. [7] developed a novel automated approach based on the text mining and machine learning for assigning a severity degree to bug reports. The proposed system was named SEVERityISsue assessment (SEVERIS). Lamkanfi et al. [6] also suggested a novel technique for severity-based bug classification. Text mining approaches were used to pre-process bug reports from Eclipse, GNOME, and Mozilla, and a Naïve Bayes classifier was then untilized. Chaturvedi et al. [8] developed the solution for determining severity degree for bug reports from NASA's project using several machine learning algorithms, i.e. SVM, RIPPER, Multinomial Naïve Bayes (MNB), Support Vector Machines (SVM, J48, K-nearest neighbor (KNN), and RIPPER. Furthermore, Tian et al. [2] introduced a novel method for recommending the priority level of bugs through the use of machine learning approaches, which is a closely related task to severity analysis. Nagwani et al. [9] proposed a method for developing taxonomy words that categorize bugs. The Latent Dirichlet Allocation (LDA) approach was employed to discover taxonomy words in their approach. Recently, Luaphol et al. [10] investigated the use of multiple machine learning algorithms (i.e. Random Forest (RF), KNN, MNB, and SVM with RBF kernel) for bug report severity prediction. The results showed that the RF algorithm combined with the *tf-icf* (term frequency-inverse class frequency) term weighting produced the most promising results.

Even though there have been several attempts to solve the problem and a number of machine learning and text mining techniques have been utilized for the aforementioned issue, the performance of the solutions may vary depending on the datasets used. Consequently, this is the purpose of our research, in which we intend to examine the outcomes of several supervised machine learning methods for severity analysis on various bug report datasets. It is noted that our study is based on a binary classification strategy in which bug reports are divided into two classes (i.e. severe class and non-severe class).

The structure of this article is as indicated below. In Sect. 2, the dataset is described in detail, whilst Sect. 3 explains the suggested method. In Sect. 4, the experimental outcomes are provided. Finally, Sect. 5 concluded the executive summary of the article.

## 2 Datasets

This study utilized four datasets, i.e. Core, Firefox, Thunderbird, and Bugzilla, that were retrieved on 10 January 2023 from https://github.com/ansymo/msr2013-bug dataset. In fact, these datasets are from the Mozilla bug tracking system. The status of these bug reports is "*closed*". This is due to the fact that bug triagers and development teams have confirmed certain bug reports as "*real-bug*" reports.

There are typically six levels of bug severity: blocker, critical, major, normal, minor, and trivial. In this study, "*real-bug*" reports with the first three severity levels were designated as severe class, whereas reports with the remaining three severity levels were designated as non-severe class.

Table 1 provides an overview of these datasets and an example of bug report can be shown as Fig. 1.

**Table 1** An overview of the datasets used

| Datasets | Number of bug reports in Severe Class | Number of bug reports in Non-severe Class | Training Set Used | | | Test Set Used |
|---|---|---|---|---|---|---|
| | | | Severe Class | Non-severe Class | Severe Class | Non-severe Class |
| Core | 15,233 | 2,931 | 2,052 | 2,052 | 879 | 879 |
| Firefox | 16,322 | 5,922 | 4,146 | 4,146 | 1,776 | 1,776 |
| Thunderbird | 4,941 | 1,867 | 1,307 | 1,307 | 560 | 560 |
| Bugzilla | 957 | 1,181 | 670 | 670 | 287 | 287 |

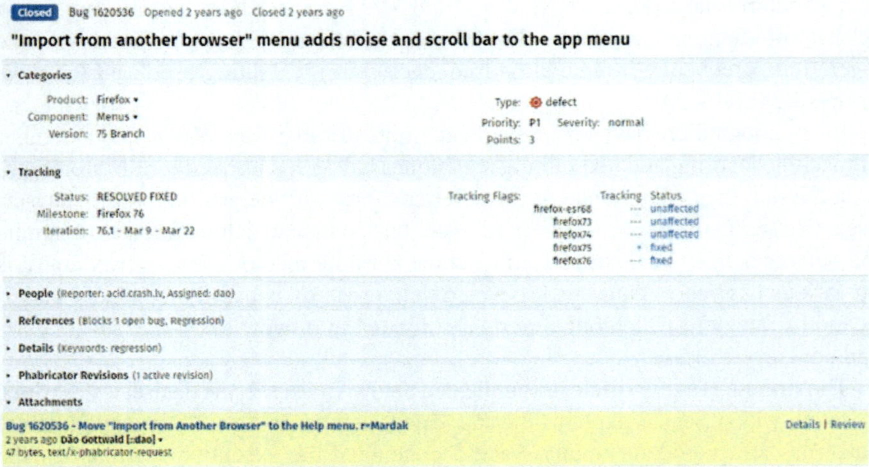


**Fig. 1** Example of bug report

Typically, the three basic components of a bug report are the title (or executive summary), the description, and the discussion. The title provides an overview of a bug report. In the meanwhile, the description provides particular details about each issue report, while the discussion provides references or comments from other end-users. However, many past studies of bug reports utilize simply the summary because it has been demonstrated that this component has less noise. Therefore, here, we also utilized only the summary section for this study.

## 3 Analysis Method for Bug Severity Identification

This section discusses the method for constructing the bug severity analyzer, which is used to determine the severity levels of bug reports. Each step of bug report pre-processing can be described in further detail below.

### *3.1 Pre-processing Bug Reports*

There are several sub-steps involved in preparing bug reports. Each stage is discussed in further detail below.

**Spelling Corrections.** Numerous bug reports contain misspellings; consequently, they must be edited to eliminate linguistic ambiguity.

**Tokenization and Punctuation Removal.** This approach was used to split text into "words" or "unigram word." After splitting text into "word" form, it requires to perform the punctuation removal process because the punctuation in the text adds noise to the sentence, causing ambiguity during training the model.

**Stop-word Removal.** Stop words are the insignificant words that should be filtered out prior to text processing. The removal of these terms helps to minimize the amount of low-level information in a text in order to give more focus on the essential information.

**CamelCase Separation.** CamelCase combines two words or abbreviations into a single word without punctuation or intervening spaces (e.g. UrlBar, AutoComplete) [11, 12]. CamelCase can help to denote the software's specificity. In order to address the issue of short text, we enhanced CamelCase features by separating them into single words; both the original CamelCase terms and their single-word counterparts are then employed as features.

**Lowercase Conversion.** Converting all the uppercase words into lowercase words allows all word strings follow a consistent format. This also helps to ambiguity during training the model.

**Word Inflection and Lemmatization.** These steps were taken to transform a unigram word into its singular form. During computerized language processing, inflections provide a range of word forms and ambiguity.

Table 2 is an illustration of bug report preprocessing.

**Table 2** An example of bug report pre-processing

| Pre-processing Step | Results |
|---|---|
| Original bug report's summary | Staff cannot use the AutoComplete function. Its work is wong |
| Spelling corrections | Staff cannot use the AutoComplete function. Its work is wrong |
| Tokenization and punctuation removal | Staff / cannot / use / the / AutoComplete / function / Its / work / is / wrong |
| Stop-word removal | Staff / cannot / use / AutoComplete / function / work / wrong |
| CamelCase separation | Staff / cannot / use / AutoComplete / Auto / Complete / function / work / wrong |
| Lowercase conversion | staff / cannot / use / autocomplete / auto / complete / function / work / wrong |
| Word inflection and lemmatization | staff / cannot / use / autocomplete / auto / complete / function / work / wrong |

## 3.2 Representation of Bug Reports and Term Weighting

The pre-processed bug reports were then represented in vector space model format (VSM), and each bug report feature (or word) was assigned a weight using a term weighting scheme. This study used *term frequency* (*tf*) as a term weighting scheme since several prior studies on bug reports found that *tf* yielded favorable findings for bug report analysis. The *tf* formula can be expressed as follows.

$$tf(t_i, d_j) = \log(1 + freq(t_i, d_j)) \tag{1}$$

where $freq(t_i, d_j)$ is the raw count of a word (or term) in a bug report.

## 3.3 Modeling of Bug Severity Analyzer

This section describes the machine learning algorithms that are utilized to model bug severity analyzer.

**Logistic Regression (LR).** This approach may be used to solve classification problems by establishing thresholds for each class's projected probability [13]. LR is often employed for binary classification, however it may be readily extended to several classes. A sigmoid function is applied to the weighted combination of the input features in LR classifiers. Using the sigmoid function, any real number may be turned into a number between 0 and 1.

**Random Forest (RF).** This algorithm is a decision tree-based ensemble learning approach using several decision trees [14]. The RF algorithm employs bagging concept and feature randomization to construct a forest of uncorrelated trees, which yields more accurate predictions than a single tree. This approach may prevent the issue of overfitting. We developed 100 decision trees for our forest as part of our research.

**Support Vector Machines (SVM).** This approach is used to partition data into classes by building a hyperplane and a decision boundary [15]. This algorithm computes the margin, which is the distance between a line and its support vectors, and identifies the support vectors, which are the points closest to the hyperplane. To maximize the margin as much as feasible is the objective. The ideal hyperplane is the hyperplane with the greatest margin. When the maximum margin is attained, the decision boundary utilized to differentiate classes is as broad as feasible, enabling classes to be distinguished more clearly. The radial basis function (RBF) kernel was utilized for the SVM approach in this study, since previous studies shown that this kernel function produced satisfactory results.

**Long Short-Term Memory (LSTM).** It is a deep learning algorithm for sequence processing. It consists of three gates: supplementary gates, input gates, and forget gates [16, 17]. The extra gates can exert greater control over the gradient, enabling the retention or discarding of certain data. These gates are sigmoid functions with outputs in the interval [0,1] that may pass either limited or complete data. A score of 0 means that the information has been fully discarded, whereas a score of 1 indicates that the information has been completely passed. In addition to the hidden layers, LSTM includes layers known as the cell state, commonly denoted $C_t$, to prevent old information from deteriorating too rapidly. LSTM is the name of the structure because it uses short-term memory processes to construct longer memories. The LSTM consists of four distinct components, i.e. input gate, forget gate, cell gate, and output gate. Each component of the LSTM is explained in detail following. Input gate is used to get new information $x^t$ using two functions (i.e. $r_t$ and $d_t$). The forget gate is denoted by $f_t$. It controls whether to maintain or delete the memory's current value. For the cell gate, the preceding cell state's input ($C_{t-1}$) is pointwise multiplied by the forget gate output. If the forget output is 0, the preceding cell output is discarded ($C_{t-1}$). To update the new cell state, this output is pointwise added to the input gate output ($C_t$). The current cell state will be sent onto the next LSTM unit. For the output gate, the hidden state is utilized for prediction and contains information about prior inputs. The output gate controls the current hidden state ($h_t$). The current input ($x_t$) and the prior hidden state ($h_{t-1}$) are sent to the sigmoid function. This output is multiplied with the tanh function's output to determine the current hidden state. The final outputs of a typical LSTM unit are the current state ($C_t$) and present hidden state ($h_t$). The LSTM architecture can be seen in Fig. 2.

We provided the algorithm setting following. The batch size remained at 32, the number of epochs remained at 20, and the dropout value was provided at 0.2. Dropout is typically utilized in deep learning to minimize overfitting. The dropout approach randomly deactivates or activates a layer's neurons throughout each cycle. As if some

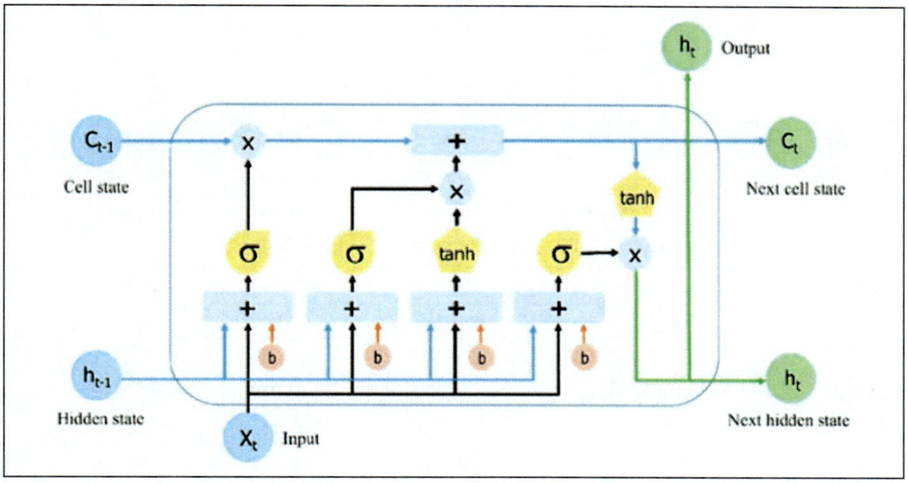


**Fig. 2** The LSTM architecture

weights were set to zero. Each iteration is utilized to enhance the model by analyzing a slightly different structure of itself. In addition, the length of the window was 5. 100 and 1 constituted the total number of neurons and densities, respectively. Tanh was the activation function, while Adam was the optimization function.

## 4 The Results

The accuracy, F1, and the Matthews Correlation Coefficient (MCC) measures were used to evaluate the performance of each approach in this study. It is noted that if a MCC value near to 1 indicates that all classes are adequately expected, despite the fact that some are disproportionately underrepresented (or overrepresented) [18]. The results of each approach can be presented in Table 3.

Table 3 showed that the LR method worked well in evaluating the severity of bug reports in the Core, Firefox, and Bugzilla datasets, whereas the LSTM technique performed well in the Thunderbird dataset.

The LR approach performed well in determining the severity level of errors discovered in bug reports because this algorithm makes no assumptions about class distributions in feature space. It not only indicates the appropriateness of a predictor (coefficient size), but also the direction of relationship (severe or non-severe). Furthermore, as the majority of text classification tasks are linearly separable, the LR approach

**Table 3** The results

| Product | Algorithms | Accuracy | F1 | MCC |
|---|---|---|---|---|
| Core | **LR** | **0.8927** | **0.7708** | **0.5495** |
| | RF | 0.8871 | 0.7170 | 0.4910 |
| | SVM with RBF | 0.8915 | 0.7342 | 0.5220 |
| | LSTM | 0.8770 | 0.7457 | 0.5024 |
| Firefox | **LR** | **0.8029** | **0.7140** | **0.4389** |
| | RF | 0.8002 | 0.6697 | 0.3981 |
| | SVM with RBF | 0.8020 | 0.6810 | 0.4085 |
| | LSTM | 0.7977 | 0.6902 | 0.4150 |
| Thunderbird | LR | 0.7834 | 0.7047 | 0.4162 |
| | RF | 0.7753 | 0.6495 | 0.3503 |
| | SVM with RBF | 0.7834 | 0.6510 | 0.3724 |
| | **LSTM** | **0.7878** | **0.7198** | **0.4439** |
| Bugzilla | **LR** | **0.7383** | **0.7367** | **0.4793** |
| | RF | 0.7407 | 0.7345 | 0.4734 |
| | SVM with RBF | 0.7290 | 0.7146 | 0.4672 |
| | LSTM | 0.7220 | 0.7196 | 0.4395 |

typically provides sufficient accuracy for many basic data sets, and it performs well when the dataset is linearly separable. It is also noted that the LR algorithm has a lower proclivity towards overfitting.

Consider the F1 and MCC scores. MCC is a statistic with a single value that summarizes the confusion matrix. It varies between -1 and 1. Typically, F1 score disregards the number of real negatives. In contrast, MCC generously extends its care to each of the confusion matrix's four entries (true positives, true negatives, false positives and false negatives). However, in this study, the MCC scores are lower than the F1 scores. It indicates that the classifier resembles a random guess classifier extremely closely. Based on these, it may be able to conclude that MCC identifies the inefficiency of the classifier in identifying negative class samples.

## 5 Conclusion and Future Work

Bug reports are an essential source of information that may be used to eliminate bugs. Determining the severity level of the reported problem is a part of bug report analysis. However, manually assigning severity to a large number of bug reports is time-consuming, and human analysis of software problem severity may be vulnerable to prejudice. Therefore, automated analysis is necessary for identifying the severity of software faults contained inside bug reports. Even though there have been several attempts to solve the problem and a variety of machine learning and text mining techniques have been implemented, the performance of the solutions may differ depending on the datasets used. As a result, this is the purpose of our research, in which we intend to evaluate the effectiveness of numerous machine learning approaches, namely Logistic Regression (LR), Random Forest (RF), Support Vector Machine (SVM) with the RBF kernel, and Long Short-Term Memory (LSTM) for severity analysis on multiple bug report datasets. Those datasets are Core, Firefox, Thunderbird, and Bugzilla that are obtained from the Mozilla bug tracking system. Our research is based on a binary classification methodology that divides bug reports into two classes: severe class and non-severe class. The experimental results demonstrated that the LR method worked well in evaluating the severity of bug reports in the Core, Firefox, and Bugzilla datasets, whereas the LSTM technique performed well in the Thunderbird dataset. In future study, we will examine different strategies, such as an ensemble method, for determining bug severity level from bug reports.

## References

1. Raymond E (1999) The cathedral and the bazaar. Knowl. Technol. Policy 3(3):23–49
2. Tian, Y., Lo, D., Sun, C.: DRONE: predicting priority of reported bugs by multi-factor analysis. In: 2013 29th IEEE International Conference on Software Maintenance (ICSM), pp. 200–209, Eindhoven, Netherlands (2013)

3. Ahsan, S.N., Ferzund, J., Wotawa, F.: Automatic software bug triage system (BTS) based on latent semantic indexing and support vector machine. In: ICSEA 2009. Fourth International Conference on Software Engineering Advances, pp. 216–221, Porto, Portugal (2009)
4. Lamkanfi, A., Demeyer, S., Giger, E., Goethals, B.: Predicting the severity of a reported bug. In: 2010 7th IEEE Working Conference on Mining Software Repositories (MSR), pp.1–10, Cape Town, South Africa (2010)
5. Defect severity classification in software testing (with an example). http://www.zyxware.com/articles/3559/defect-severity-classification-insoftware-testing-with-an-example. Accessed 24 May 2022
6. Cubranic, D., Murphy, G.C.: Automatic bug triage using text categorization. In: Proceedings of the Sixteenth International Conference on Software Engineering & Knowledge Engineering (2004)
7. Menzies, T., Andrian, M.: Automated severity assessment of software defect reports. In: IEEE International Conference on Software Maintenance, pp. 346–355, Beijing, China (2008)
8. Chaturvedi, K.K., Singh, V.B.: Determining bug severity using machine learning techniques. In: International Conference on Software Engineering (CONSEG), pp. 1–6, Indore, India (2012)
9. Nagwani, N.K., Verma, S., Mehta, K.K.: Generating taxonomic terms for software bug classification by utilizing topic models based on Latent Dirichlet Allocation. In: International Conference on ICT and Knowledge Engineering (ICT&KE), Bangkok, Thailand (2013)
10. Luaphol B, Polpinij J, Kaneampornpun J (2022) Text mining approaches for dependent bug report assembly and severity prediction. Int. Arab J. Inf. Technol. 19(6):916–924
11. Antoniol, G., Ayari, K., Di Penta, M., Khomh, F., Guéhéneuc, Y.G.: Is it a bug or an enhancement? A text-based approach to classify change requests. In: Proceedings of the 2008 conference of the center for advanced studies on collaborative research: meeting of minds, pp. 304–318 (2008)
12. Luaphol, B., Srikudkao, B., Kachai, T., Srikanjanapert, N., Polpinij, J.: Feature comparison for automatic bug report classification. In: International Conference on Computing and Information Technology, pp. 69–78 (2019)
13. Hassan SU, Ahamed J, Ahmad K (2022) Analytics of machine learning-based algorithms for text classification. Sustain. Oper. Comput. 3:238–248
14. Islam, M.Z., Liu, J., Li, J., Liu, L., Kang, W.: A semantics aware random forest for text classification. In: Proceedings of the 28th ACM International Conference on Information and Knowledge Management, pp. 1061–1070 (2019)
15. Joachims, T.: Text categorization with support vector machines: learning with many relevant features. In: European Conference on Machine Learning, pp. 137–142 (1998)
16. Hochreiter S, Schmidhuber J (1997) Long Short-Term Memory. Neural Comput. 9(8):1735–1780
17. Wu, C.H., Lu, C.C., Ma, Y.F., Lu, R.S.: A new forecasting framework for bitcoin price with LSTM. In: 2018 IEEE International Conference on Data Mining Workshops (ICDMW), Singapore (2018)
18. Ramay WY, Umer Q, Yin XC, Zhu C, Illahi I (2019) Deep neural network-based severity prediction of bug reports. IEEE Access 7:46846–46857

# Sliding-Window Technique for Enhancing Prediction of Forex Rates

**Siranee Nuchitprasitchai, Orawan Chantarakasemchit, and Yuenyong Nilsiam**

**Abstract** International businesses are using foreign currencies as part of their activities. To optimize the cost of foreign exchange, the prediction of the exchange rate can be crucial. For time series forecasting, moving average techniques should be adopted. In this study, the exchange rate of EUR/USD is mainly focused on. The data under consideration are the Forex transaction rates which include open, high, low, and close. Furthermore, four financial factors are the dollar index, interest rate, inflation rate, and real growth of domestic products which are also included as inputs. Four models were employed: linear regression, multilayer perceptron, recurrent neural networks, and general regression neuron networks for predicting the Forex rate. Mean squared error is used for comparing forecasting performance. The results show that applying a sliding-window and moving average techniques improves the efficiency of the prediction.

**Keywords** FOREX rates · Dollar Index · Interest Rate · Inflation Rate · Gross Domestic Product · Moving Average Techniques · Sliding Window Techniques · Deep Learning

## 1 Introduction

In international trading, buyers and sellers should be able to exchange different currencies. The foreign exchange market (Forex) is where buying, selling, and exchanging currencies occur. Normally, foreign exchange rates are in pairs of currencies, e.g., Euro to U.S. dollar (EUR/USD).

The highest foreign exchange trading value per day reached $6.6 trillion dollars in April 2019 [1]. The market has grown by about 30 percent since 2016. There have been many new investors and speculators entering the foreign exchange market during the last few years. However, to leverage the foreign exchange market, the

S. Nuchitprasitchai (✉) · O. Chantarakasemchit · Y. Nilsiam
King Mongkut's University of Technology North Bangkok, Bangkok 10800, Thailand
e-mail: siranee.n@itd.kmutnb.ac.th

 
P. Meesad et al. (eds.), *Proceedings of the 19th International Conference on Computing and Information Technology (IC2IT 2023)*, Lecture Notes in Networks and Systems 679,
https://doi.org/10.1007/978-3-031-30474-3_18

directions of exchange rates need to be known in advance. Therefore, accuracy of prediction of an exchange rate of interest is useful for business, investing, and trading.

There are many techniques for input data preparation and creating a prediction model. In previous research, four factors were found that are highly related to the Forex rates, so those factors are necessary to be used as input for a prediction model [2]. The forex rate is also an input data. A simple moving average (SMA) can be applied to the input dataset which helps to improve the efficiency of prediction models, the multilayer perceptron (MLP) and the linear regression model (LM). Another study [3], which weighted the moving average (WMA) and the exponential moving average (EMA) were adopted to provide the same input. A model called recurrent neuron networks (RNNs) was added to the study. The result shows that applying moving average techniques to the dataset improves the accuracy of the predictions on average.

In this research, a sliding-window technique is employed to the forex data. Four models were adopted: the linear regression model (LM), the multilayer perceptron (MLP), recurrent neural networks (RNNs) and general regression neural networks (GRNNs). The performance of the predictions based on the input data and models is analyzed using mean squared error (MSE) and its average value.

## 2 Literature Review

A sliding-window model was used to enhance data sampling and applied to many other uses [4]. Two types of sliding window models are fixed size and timestamp-based windows. Using the sliding-window technique can cause some drawbacks, namely, are extra costs and nondeterministic use of memory. The optimization of data sampling can be achieved by utilizing equivalent-width partitions, covering decomposition, generating implicit events, and black-box reduction from sampling without replacement to sampling with replacement.

A sliding-window technique was used to improve a forecast system of the foreign exchange rate [5]. The research shows that the proposed system called a sliding-window metaheuristic optimization-based forecast (SMOF) has very low errors. The percentage of mean absolute errors are 0.2532% for the Canadian dollar-USD exchange rate (CAN/USD) and 0.169% for the Euro-USD exchange rates (EUR/USD). The high volume of data of the foreign exchange market can cause problems for building a forecasting model. The sliding-window technique is employed to help secure the impact of new data to a training model because it weights the latest data and ignores the most aged data. The method also eliminates the problem of having too much data to operate on due to the fixed size of the sliding-window. Another study in economics and finance examined exchange rate patterns using autocorrelation-based transformation [6]. The results indicate that adjusting the length of the sliding window properly can improve forecasting accuracy.

In the previous work [2], some basic techniques were presented, which can be summarized in the following paragraphs.

A simple moving average is a statistical technique used to analyze time-series data and to identify trends and patterns in a dataset. It is a commonly used approach in technical analysis and is particularly useful in financial markets for smoothing out price data and identifying trends. To calculate a simple moving average, a certain number of data points are averaged together, with each data point given equal weight. For example, to calculate a 3-day simple moving average of an exchange rate, the exchange rates for the last 3 days would be added together and then divided by 3. The formula can be shown as follows:

$$SMA = \frac{A_1 + A_2 + \cdots + A_n}{n} \tag{1}$$

where $n$ is a number of the whole period, and $A_n$ is value at the period $n$.

A linear regression model is a statistical method that models the relationship between a dependent variable and one or more independent variables. The model assumes a linear relationship between the variables, where changes in the independent variables are associated with corresponding changes in the dependent variable. The model estimates the slope and intercept of the line that best fits the data. The formula can be shown as follows:

$$y = \theta_1 + \theta_2 x \tag{2}$$

where $y$ is a dependent variable, $x$ is an independent variable, $\theta_1$ is an intercept value, and $\theta_2$ is the slope estimated from the data.

A multilayer perceptron (MLP) is a type of artificial neural network that is designed to learn complex nonlinear relationships between inputs and outputs. It consists of an input layer, one or more hidden layers, and an output layer. Each layer contains multiple nodes or neurons that perform mathematical operations on the input data. In an MLP, the input data is fed into the input layer, where it is transformed and passed to the next layer. Each subsequent hidden layer then applies a set of weights and biases to the output of the previous layer and applies an activation function to produce a new set of outputs. The final output layer produces the predicted output based on the learned weights and biases of the preceding layers.

The recurrent neural networks (RNNs) are a class of neural networks suitable for modeling time series or natural language. The output of each input provides feedback as an input and it is preserved as a hidden state. A formular of hidden state (h) at the point of time (3) is shown in the following model [7].

$$h(t) = g(h)(W(h)[h(t-1), x(t)] + b(h) \tag{3}$$

where h(t) is the hidden state at the point of time, g(h) is the hidden activation function, W(h) is a function of the hidden weight matrix, h(t-1) is the previous time step's hidden state, x(t) is an input vector, and b(h) is the bias.

The most common activation functions used in RNNs are the logistic sigmoid, the hyperbolic tangent, and the rectified linear unit [8].

In a sigmoid function, input will be normalized to values of 0.0 to 1.0 range [9]. The range of hyperbolic tangent function is –1 to 1. The rectified linear unit is the simplest, as it returns input directly if it is greater than 0.0 and returns 0.0 for the rest.

The general steps of RNNs [10] are as follows, with the first step initializing weight matrices with random distribution and bias with zero. The second step is computing predictions from forward propagation. The third step is computing the loss. The next step is computing gradients with backpropagation. The fifth step is updating the weights based on gradients and then to repeat the second to the fifth step.

As the output is a function of the weights and biases of all the connections, so by computing the gradient of the loss function with consideration of the network's weights obtains the information needed to train the network. Through the process of RNNs, RNNs tend to run into two problems, known as exploding gradients and vanishing gradients [11]. Exploding gradients occur when the gradient is too large as the result of creating an unstable model. Vanishing gradients occur when the gradient is too small and continue to become smaller as the result of updating the weight parameters until they become insignificant. When these problems occur, the algorithm is no longer learning.

Exploding the gradient can be fixed with a gradient clipping technique that is used to cap the maximum value for the gradient. While the vanishing gradient limitation is fixed by various networks, such as Long Short-Term Memory (LSTM) and Gated Recurrent Units (GRUs). The LSTM is a popular RNN architecture, which was introduced as a solution to the vanishing gradient problem by using the cell state. The cell state allows the LSTM to remember dependencies through time and facilitates bridging long-time lags. The cell state contains no activation functions but only linear operations for preventing the vanishing gradient problem.

The General Regression Neural Networks (GRNNs) [12] is a neural network that provides estimates of continuous variables and converges to the underlying (linear/nonlinear) regression form. The GRNNs are similar in form to the Probabilistic Neural Network (PNN) but GRNNs can be used to estimate the values of continuous dependent variables.

The calculations performed in each pattern neuron of GRNNs [13] are exp $(-D2(\mathrm{j})/2\sigma 2)$, which is the normal distribution centered in each training sample. The signals of the pattern neuron $i$, which enter into the denominator neuron are weighted with the corresponding values of the training samples, $Y(i)$. The weights on the signals entering the Numerator Neuron are one.

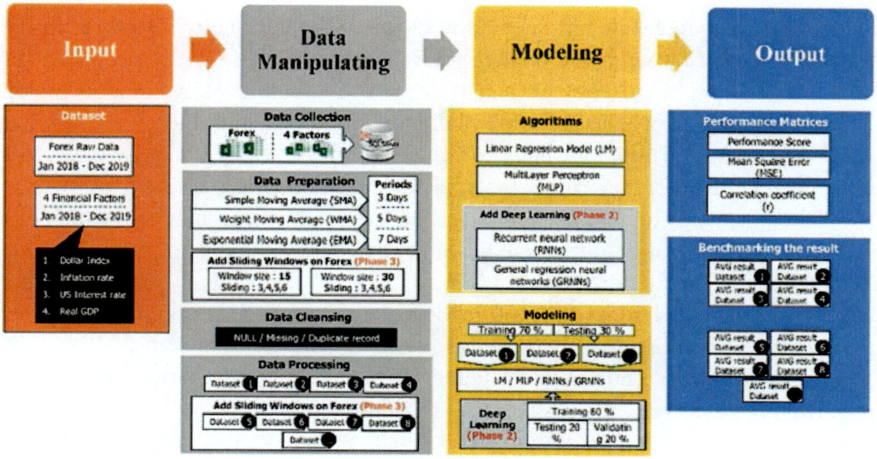


**Fig. 1** The workflow model

# 3 Methodology

## *3.1 Overview*

There are two main processes, data manipulation and modeling as shown in Fig. 1. Overall, research can be divided into three phases. The first phase focusses on selecting factors from many parameters that could affect the forex rate based on the correlation coefficient. The algorithms for modeling are LM and MLP. The second phase, moving average methods and deep learning algorithms, RNNs and GRNNs, are added to the experiment. The third phase, which is the sliding window technique is applied to the exchange rate data. The performance of different input data and models are analyzed based on mean squared error (MSE) and performance score.

Python is a high-level programming language with many useful libraries. It is used in this research for modeling. Another language which is used is called SQL for manipulating data. Many libraries in Python are used. For example, pandas which is a library for data analysis, *NumPy* is a library for complicated mathematical functions, and *Sklearn* is a library for machine learning capability of classification, regression, clustering, dimensionality reduction, and preprocessing algorithms. The deep learning libraries in Python are provided by TensorFlow.

## *3.2 Input Data*

The currency pair of Euro to U.S. Dollar (EUR/USD) was used in this research (histdata.com). There were five features in the dataset: date-time stamp, open bid quote, high bid quote, low bid quote, and close bid quote. The five features in the dataset have been chosen based on their relevance and potential predictive power for modeling changes in currency exchange rates. The date-time stamp is a necessary feature for time-series analysis, while the open bid quote, high bid quote, low bid quote, and close bid quote provide information on the trading activity and price movements of the currency pair being analyzed. These features are commonly used in technical analysis to identify trends and patterns in financial markets, which can be useful for forecasting future price movements. The exchange rates were recorded every minute in 2018 and 2019: 372,607 and 354,773, respectively. Exchange rate data is recorded every minute to capture the fast changes in currency exchange rates in the foreign exchange market, which is highly dynamic and operates 24 h a day. Therefore, analysts can gain a more detailed and nuanced understanding of the market trends and price movements.

The four economic factors of the United States of America were Dollar Index (investing.com), Interest Rate (federalreserve.gov), Inflation Rate (inflation-data.com), and Real Gross Domestic Product (GDP) (apps.bea.gov) were included to enhance the input. During 2018 and 2019, the Dollar Index was recorded daily with 261 and 269, respectively. The interest rate was recorded when there were changes in the economic policy, which occurred 16 times and 13 times, respectively. The inflation rate was recorded monthly. GDP was recorded quarterly.

### *3.3 Data Manipulation*

With regard to the data collection, all data were imported into an SQL database server.

Data preparation, SMA, WMA, EMA, and the sliding window technique were applied to the Forex dataset, except for the date-time stamp, which was expected to improve the overall performance. Each moving average was calculated in a period of 3, 5, and 7 days. For the sliding-window technique, the periods were 15 and 30 days and the sliding widths were 3, 4, 5, and 6 days. Figure 2 shows an example of the sliding-window technique. In this research, the sliding-window technique was applied according to the availability of data for creating realistic models. Models cannot be formulated in a short time.

Data cleansing, duplicate records, null values, and missing values were used.

With regard to the data processing, the two groups of datasets were the normal Forex dataset and the dataset of the applied sliding window with the Forex dataset. All datasets are shown in Table 1.

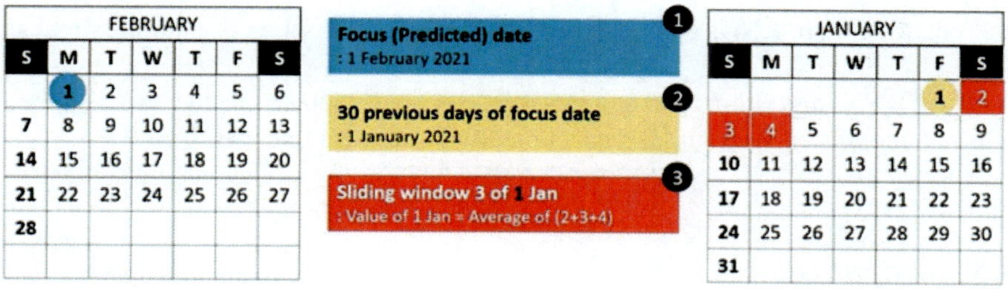


**Fig. 2** An example of the sliding-window technique

**Table 1** Dataset Details

| Dataset No | Dataset Details | | |
|---|---|---|---|
| | Group 1: The normal Forex data | | |
| 1 | Forex only | | |
| 2 | Forex with 4 factors | | |
| 3 | Forex with series of MAs | | |
| 4 | Forex with 4 factors and series of MAs | | |
| Dataset No | Dataset Details | | |
| | Group 2: The applied sliding window on Forex data | | |
| 5–8 | S3, S4, S5, S6 | W15 | |
| 9–12 | S3, S4, S5, S6 | W15 | With 4 factors and series of MAs |
| 13–16 | S3, S4, S5, S6 | W30 | |
| 17–20 | S3, S4, S5, S6 | W30 | With 4 factors and series of MAs |

* Note: MAs are moving averages, S is sliding width (3–6 days), and W is training periods (15, 30 days)

## *3.4 Modeling*

Each dataset in the data manipulation process was divided into a training dataset (70%) and a testing dataset (30%) for LM, MLP, and GRNNs algorithms. For the RNNs algorithm, each dataset was divided into three parts, training, testing, and validating datasets. The ratio was 60:20:20 percent. Each model was executed four times, one for each output. The four outputs were Open, High, Low, and Close Rate.

As for the MLP parameters, the number of neurons in the input layer was defined by the number of features in the running dataset. The Forex dataset had four features, open, high, low, and close rate. Each economic factor was a feature. Each period of each moving average was a feature. The hidden layer was 50 layers and 10 neurons for each. The activation function for the hidden layer was the rectified linear unit function. The output layer had one neuron for output.

With regard to the RNNs algorithm, the "hyperbolic tangent" was used as the function activation and "sigmoid" was used as the recurrent activation, one batch size, and one time-step. RNNs can be used for modeling time series or natural language [7].

The GRNNs algorithm was for a small dataset, so the datasets needed to be divided into two months each. A larger dataset could not be executed [12].

### 3.5 *Output Data*

The performance metrics of each input dataset and algorithm were used to evaluate the efficiency. The performance score was based on the prediction errors, MSE, so the least is the best.

## 4 Results

The sliding window technique was applied to the Forex data using a training period (W) of 15 and 30 days and a sliding size (S) of 3, 4, 5, and 6 days. For GRNNs, a training period is two months, for example, M1M2 represented January and February. Therefore, each combination was a new dataset in a total of eight. The MSE of sliding window data with 4 factors and MAs for the models, LM, MLP, RNNs, and GRNNs, are presented in Tables 2, 3, 4, and 5.

**Table 2** MSE of Sliding Window Data with 4 Factors and MAs for LM

| Dataset | Mean Square Error (MSE) (Represented: Result * $10^7$) | | | | | |
|---|---|---|---|---|---|---|
| | Open | High | Low | Close | Average | Rank |
| **W15** | | | | | | |
| S3W15 | 39.19454 | 39.20452 | 39.17201 | 39.16159 | **39.18317** | #2 |
| S4W15 | 39.29534 | 39.31333 | 39.29185 | 39.28496 | **39.29637** | #4 |
| S5W15 | 39.34797 | 39.37818 | 39.34829 | 39.35047 | 39.35623 | |
| S6W15 | 38.95920 | 38.97888 | 38.96114 | 38.95748 | **38.96417** | #1 |
| **W30** | | | | | | |
| S3W30 | 39.35005 | 39.36805 | 39.35373 | 39.34688 | 39.35468 | |
| S4W30 | 39.21595 | 39.23182 | 39.20798 | 39.20493 | **39.21517** | #3 |
| S5W30 | 39.30641 | 39.31647 | 39.30461 | 39.29455 | 39.30550 | |
| S6W30 | 39.30584 | 39.31630 | 39.30334 | 39.29419 | **39.30492** | #5 |

Note: The bold text represents the best value

**Table 3** MSE of Sliding Window Data with 4 Factors and MAs for MLP

| Dataset | Mean Square Error (MSE) (Represented: Result * $10^7$) | | | | | |
|---|---|---|---|---|---|---|
| | Open | High | Low | Close | Average | Rank |
| **W15** | | | | | | |
| S3W15 | 775.7683 | 241.5309 | 246.7861 | 1341.074 | 651.2899 | |
| S4W15 | 210.2475 | 183.9684 | 1283.458 | 712.1530 | **597.4569** | #4 |
| S5W15 | 465.0992 | 117.1664 | 146.8848 | 164.1712 | **223.3304** | #1 |
| S6W15 | 241.7604 | 479.9975 | 570.2007 | 664.7139 | **489.1681** | #2 |
| **W30** | | | | | | |
| S3W30 | 1278.099 | 487.2580 | 183.7486 | 872.7622 | 705.4671 | |
| S4W30 | 1508.686 | 204.7256 | 452.4431 | 145.2749 | **577.7826** | #3 |
| S5W30 | 946.7523 | 1085.120 | 197.6228 | 161.2350 | **597.6827** | #5 |
| S6W30 | 1418.834 | 1097.098 | 216.6718 | 253.1366 | 746.4353 | |

Note: The bold text represents the best value

**Table 4** MSE of Sliding Window Data with 4 Factors and MAs for RNNs

| Dataset | Mean Square Error (MSE) (Represented: Result * $10^7$) | | | | | |
|---|---|---|---|---|---|---|
| | Open | High | Low | Close | Average | Rank |
| **W15** | 9989.1 | 32,935.3 | 15,738.8 | 11,059.9 | 17,430.8 | |
| S3W15 | 28,010.3 | 26,968.6 | 11,907.1 | 8984.0 | 18,967.5 | |
| S4W15 | 5590.7 | 3684.0 | 6261.3 | 3329.6 | **4716.4** | #1 |
| S5W15 | 4961.5 | 19,804.4 | 4544.7 | 3463.3 | **8193.5** | #2 |
| S6W15 | 21,612.8 | 25,526.8 | 9108.1 | 14,076.9 | **17,581.2** | #5 |
| **W30** | 62,624.4 | 5567.4 | 40,701.2 | 17,017.7 | **31,477.7** | |
| S3W30 | 28,191.3 | 24,848.3 | 18,051.5 | 12,313.7 | 20,851.2 | |
| S4W30 | 4492.7 | 17,077.3 | 27,709.3 | 20,230.9 | **17,377.6** | #4 |
| S5W30 | 27,942.1 | 24,073.7 | 4293.8 | 7557.4 | **15,966.8** | #3 |
| S6W30 | 29,337.5 | 7139.1 | 29,298.9 | 23,856.3 | 22,408.0 | |

Note: The bold text represents the best value

**Table 5** MSE of Sliding Window Data (S4W15) with 4 Factors and MAs for GRNNs

| Dataset | Mean Square Error (MSE) | | | | |
|---|---|---|---|---|---|
| | Open | High | Low | Close | Year |
| M1M2 | **0.026662** | **0.026763** | **0.026639** | **0.026658** | 2018 |
| M3M4 | 0.079281 | 0.078629 | 0.078819 | 0.079293 | |
| M5M6 | 0.034385 | 0.034895 | 0.034209 | 0.034393 | |
| M7M8 | 0.114142 | 0.113958 | 0.114402 | 0.114201 | |
| M9M10 | 0.056101 | 0.056035 | 0.056437 | 0.056633 | |
| M11M12 | 0.043050 | 0.043064 | 0.043044 | 0.042974 | |
| Dataset | Mean Square Error (MSE) | | | | |
| | Open | High | Low | Close | Year |
| M1M2 | 0.032493 | 0.033761 | 0.032557 | 0.032455 | 2019 |
| M3M4 | **0.026344** | **0.026232** | **0.026325** | **0.026084** | |
| M5M6 | 0.069654 | 0.071163 | 0.068586 | 0.069679 | |
| M7M8 | 0.023838 | 0.023610 | 0.023927 | 0.023877 | |
| M9M10 | 0.060207 | 0.060115 | 0.059876 | 0.060239 | |
| M11M12 | 0.045428 | 0.045634 | 0.045485 | 0.045445 | |

Note: The bold text represents the best value

## 5 Conclusion

Based on the results in the previous section, the best dataset for LM is S6W15, for MLP is S5W15, and for RNNs is S4W15. The best dataset for GRNNs is M1M2 for 2018 and M3M4 for 2019. When comparing the accuracy between the models, the most effective model is LM. The availability of training data is crucial, so the sliding-window technique is needed. Overall, the sliding-window technique improves the prediction performance of the models. However, the training period and sliding width are variables which need to be fine-tuned to obtain the optimal results.

In future work, the models for machine learning should be enhanced by selecting proper variables and adjusting parameters for the models. This process should be done automatically by software.

From Table 4, the best dataset is S4W15, so it was used to create a model for GRNNs.

## References

1. Foreign exchange turnover in April 2019. https://www.bis.org/statistics/rpfx19_fx.htm. Accessed 20 Jan 2020
2. Chantarakasemchit, O., Nuchitprasitchai, S., Nilsiam, Y.: Forex rates prediction on EUR/USD

with simple moving average technique and financial factors. In: 17th International Conference on Electrical Engineering/Electronics, Computer, Telecommunications and Information Technology (ECTI-CON), pp. 771–774. IEEE (2020)
3. Chantarakasemchit, O., Nuchitprasitchai, S.: Enhancing forex rates prediction with machine learning on EUR to USD with moving average methods and financial factors. In: Recent Advances in Information and Communication Technology on Proceedings of the 17th International Conference on Computing and Information Technology (IC2IT 2021), pp. 44–54. Cham: Springer International Publishing (2021)
4. Braverman, V., Ostrovsky, R., Zaniolo, C.: Optimal sampling from sliding windows. In: Proceedings of the twenty-eighth ACM SIGMOD-SIGACT-SIGART symposium on Principles of database systems, pp. 147–156. (2009)
5. Chou JS, Truong TTH (2019) Sliding-window metaheuristic optimization-based forecast system for foreign exchange analysis. Soft Comput. 23(10):3545–3561
6. Yao CZ, Lin QW (2017) Recurrence plots analysis of the CNY exchange markets based on phase space reconstruction. North Am. J. Econ. Financ. 42:584–596
7. Dautel AJ, Härdle WK, Lessmann S, Seow HV (2020) Forex exchange rate forecasting using deep recurrent neural networks. Digital Financ. 2:69–96
8. Amidi, A., Amidi, S.: Recurrent Neural Networks Cheatsheet, Stanford University. https://stanford.edu/~shervine/teaching/cs-230/cheatsheet-recurrent-neural-networks. Accessed 20 Mar 2021
9. Brownlee, J.: A gentle introduction to the rectified linear unit (ReLU). Machine learning mastery, 6. https://machinelearningmastery.com/rectified-linear-activation-function-for-deep-learning-neural-networks/. Accessed 20 Mar 2021
10. Nabi, J.: Recurrent neural networks (RNNs). Medium. https://towardsdatascience.com/recurrent-neural-networks-rnns-3f06d7653a85. Accessed 20 Mar 2021
11. IBM Cloud Education, What are Recurrent Neural Networks? https://www.ibm.com/cloud/learn/recurrent-neural-networks. Accessed 20 Mar 2021
12. Specht DF (1991) A general regression neural network. IEEE Trans Neural Netw. 2(6):568–576
13. General Regression Neural Network (GRNN). https://minds.wisconsin.edu/bitstream/handle/1793/7779/ch2.pdf?sequence=14. Accessed 20 Mar 2021

# Author Index





P. Meesad et al. (eds.), *Proceedings of the 19th International Conference on Computing and Information Technology (IC2IT 2023)*, Lecture Notes in Networks and Systems 679,
https://doi.org/10.1007/978-3-031-30474-3